Biosemiotics

Volume 12

Combining research approaches from biology, semiotics, philosophy and linguistics, the field of biosemiotics studies semiotic processes as they occur in and among living systems. This has important implications and applications for issues ranging from natural selection to animal behaviour and human psychology, leaving biosemiotics at the cutting edge of the research on the fundamentals of life.

The Springer book series *Biosemiotics* draws together contributions from leading scholars in international biosemiotics, producing an unparalleled series that will appeal to all those interested in the origins and evolution of life, including molecular and evolutionary biologists, ecologists, anthropologists, psychologists, philosophers and historians of science, linguists, semioticians and researchers in artificial life, information theory and communication technology.

Anton Markoš • Jan Horský

The Seeds of Forms and the Forms of Seeds

Narratives in Historiography and in Evolutionary Biology

Springer

Anton Markoš
Philosophy and History of Sciences
Charles University Prague
Prague, Czech Republic

Jan Horský
History
Charles University, Faculty of Humanities
Prague, Praha, Czech Republic

ISSN 1875-4651 ISSN 1875-466X (electronic)
Biosemiotics
ISBN 978-3-032-16130-7 ISBN 978-3-032-16131-4 (eBook)
https://doi.org/10.1007/978-3-032-16131-4

This Springer imprint is published by the registered company Springer Nature Switzerland AG
The registered company address is: Gewerbestrasse 11, 6330 Cham, Switzerland

Deep Memories

Underneath, in augmented note values, as a cantus firmus, *an ancient Gregorian chant (a chant that goes back to an immemorial and probably non-European past); above it, in shorter note values, unfolds the polyphonic accompaniment's melody. This embrace of two melodies belonging to two different eras (centuries apart) has something marvelous about it: like reality and parable at once, here is the birth of European music as art: a melody is created to go in counterpoint with another, very old, melody whose origins are almost unknown.*

Milan Kundera 1993

In other words, Priscilla and I are only meeting places for messages from the past: not only for messages among themselves, but for messages meeting answers to messages. And as the different elements and molecules answer messages in different ways—imperceptibly or boundlessly different—so the messages vary according to the world that receives them and interprets them, or else, to remain the same, they are forced to change. You might say, then, that the messages are not messages at all, that a past to transmit doesn't exist, and only so many futures exist which correct the course of the past, which give it form, which invent it.

Italo Calvino 1976

Introduction

Memory and experience. The principal features of life. All life! Memory of recent events in our lives and/or the life of our community; memory of life's artifacts, as coral reefs, triumphal arches or road networks; of biological membranes and other very conservative cell structures and procedures. Add paleontological findings; glacial periods or several months long droughts; knowledge of idioms and courtesy phrases and their appropriate usage at the given time…; and before all, memory in a form of coded messages – be it genetic texts, bird songs, hormones and pheromones, or texts written by humans such as books or burial inscriptions.

Experience concerns the extraction of meaning—here and now—from all those inputs coming from the outside as they are confronted by the internal settings (memory) of an individual, of a community, and/or a lineage. In short: experience organizes ways of living by negotiating ecosystems of relations to construct a narrative of how the world works.

History took place and is no longer accessible—save scattered, scrappy fragments of memory; scattered through—often vast—expanses of time and space. To assess their meaning—meaning for us, living here and now! —we, living beings, must combine such traces into coherent, unbroken chains that allow the construction of a plausible storyline of events past. Here we—authors—met: JH is a historiographer experienced in methods of interpreting the histories of human societies; AM is a biologist interested in evolution and the coexistence of non-human living beings. In this book, we tried "zooming out" from human affairs to all of life.

Zooming out to all of life (while simultaneously excluding all that is non-life) requires a working definition of life acceptable to both authors: first, we must rid ourselves of the classical divide "nature vs. culture" that delineates a boundary between humans and the rest of life. (Well, some "higher" animals may be grudgingly accepted as "cultural", but we seek to welcome plants, bacteria and other life forms to the party.) We dare to do so, and seek common denominators that may serve both the cellular nature of all living beings as well as their gregarious character that organizes the cells into communities (and these into higher-rank communities), be it coral reefs, religious communes, bacterial mats, nations and languages, rituals, etc. At a closer look, the two life categories differ from each other as to the way they come into existence: cells are born from other cells in an uninterrupted chain since life's origins, i.e., all life is interconnected through the common descent of all cells. Cell and supracellular communities, on the other hand, are based on the cooperation of different kinds of cells; some may also come into existence at the dawn of life, but they can easily be formed de novo and fade away again. To avoid confusions with terminology, we decided to coin new concepts: "Closure 1" for the lasting cellular forms, and "Closure 2" for communities, making use of the archaic meaning of "closure" as a means of enclosing.

"Life and semiosis are coextensive" is a definition of life as coined by biosemioticians. All living beings that happen to live in any given present, are busy constructing the world and negotiating with their contemporaries the ways of their living. This book is an attempt to light up one possible scenario of this astonishing process that we—living beings—both create and are creations of.

The aim of the first chapter is to bring the reader closer to the relationship between the data (sources, measurement results, etc.) and the resulting narratological achievement—the story. We would like to draw her or him into the whirlwind of things with the second chapter where we analyze a short story *The Fall of the House of Voticky* (from Čapek, 1994).

The third chapter will introduce the basics of semiotics, that is, the art of interpretation especially as performed by John Deely and Umberto Eco. The chapter acquaints the reader with the terminology and the way of grasping the world within the framework of semiotic doctrine. Semiotics was created to serve the needs of human culture; the biosemiotic endeavor seeks to extend it to all living things—more on this in the chapter.

The fourth chapter is a kind of theoretical training that takes us into the ways in which narrative arises in the biological and historical sciences as we living creatures interpret, hence construct, the world. Here we focus on human affairs and discuss the approaches to interpreting the history of the world offered by language, writing, memory, history etc., i.e., the connective structures that hold our world (or parts of it) together; again, with hints toward biosemiotic broadeneing of the topic in mind.

The next three chapters continue in this vein, each time analyzing different attempts to interpret history—including human history, but with digressions into issues of the evolution of life. Chapter 5 focuses on semiotics and its interpretative schemes; the reader can trace the evolution of understandings of historical narratives and interpretations of canonical themes that are deeply embedded in culture. In Chap. 6 we will see how a community's self-understanding changes over different time horizons, separated by several generations. This is followed by Chap. 7, which introduces us to historical "generalities" and illustrtes them on a simple question from early Czech history: did slavery exist in Bohemia?

The rest of the book is more concerned with the history of life in general. Chapter will highlight some methods of reconstructing evolutionary events, while the ninth will push history somewhat aside and provide insight into "timeless" scientific methods of examining life functions. The Chaps. 10 and 11 is a summary of all this and suggests some connections with other historical models.

Many thanks to colleagues, life partners, and other friends and sympathizers for reading versions of this text and providing invaluable comments: Tereza Liepoldová, Jana Švorcová, Dina Novotná Obeidová, Lukáš Zámečník, Vít Gvoždiak, František Baluška, and Jan Toman; in addition, Fatima Cvrčková also helped with the Figures. Lucie Nováková and Gerald R. Ostdiek helped to put the text into English, and unknown rewievers help to improve parts of the text. And of course, we thank our *alma mater* Charles University for sustaining our research.

Prague, Czech Republic
November 2025

Jan Horský
Anton Markoš

Acknowledgements

We thank our *alma mater* Charles University for sustaining our research.

Competing Interests The authors have no competing interests to declare that are relevant to the content of this manuscript.

Contents

Chapter 1
Historiography and Evolutionary Biology

Abstract Introductory remarks concerning our endeavor to show parallels between history and evolution of life—and between approaches of historiography and evolutionary biology. After brief discussion of physical and hylozoical definitions, this brings us to characterization of life as an entity *always* consisting of cells. Only life can distinguish some information on the background of its environment, and to assign it a *meaning* (according to circumstations). Life comes in two basic forms, Closure 1 (cells) and Closure 2 (supracellular teams), and of the communicative network binding them together (sybiotically) within the biosphere. What follows is basic introduction into memory, basically script and structure, their interpretation by the living, and historical changes of ways of both storage of memory, and interpretation of it by using (bio)semiotic methods. We will also touch the question whether evolutionary biology does belong to the realm of biology defined as modern *science*. Would not methods developped in historical sciences help deeper understanding? Are analogies drawn between historiography and evolutionary biology allowed? If so, may they be useful at all? Finally, there comes a concise outlay of the whole book, in form of three "basic" or "semiotic" statements.

Keywords History of life and human history · Life in the universe · Life and non-life · Closures · (Bio)semiotic method

> If reality was not necessarily clear, then by implication it was right to proceed by building up knowledge of the whole from the parts, using the conjectural paradigm [...] and this was in fact the usual approach in a number of spheres of activity: physicians, historians, politicians, potters, joiners, mariners, hunters, fishermen, and women in general—were held, among others, to be adept in the vast areas of conjectural knowledge.
>
> Carlo Ginzburg & Anna Davin, 1980: 15.

A. Markoš, J. Horský, *The Seeds of Forms and the Forms of Seeds*, Biosemiotics 12, https://doi.org/10.1007/978-3-032-16131-4_1

In this book, we consider various analogies and parables that might inspire research in two major areas of historical inquiry: on the one hand, the history of human communities, cultures and civilizations, and on the other, evolutionary biology as the history of life on Earth. We will question the extent to which the paths of biological evolution can be applied to the interpretation of the development of culture (its evolution and diffusion). And vice versa: humans certainly interpret their history, but could not the methodology of historical studies enrich research within evolutionary biology? Moreover, a third question comes to mind: do different life forms also interpret (represent) their past in this way—independently of what human researchers think? With this chapter we would like to draw the reader into the midst of a maelstrom in which we will together look for—and perhaps find—some order.[1]

We are witnessing extreme statements—in biology and beyond—about what life actually is. One margin is represented by absolute denial: according to some, it is a completely useless concept that only obscures the fact that in our environment we observe a seamless continuum from physical and chemical processes to life processes, while these "higher" functions (i.e., manifestations of what we normatively call "life") are entirely explicable, even derivable, from relatively simple physical principles (chemistry included). Today's biology considers itself a natural science,[2] and so implicitly assumes the same, its task being the study of life with the aim of discovering these laws:

> So why should we be surprised if quantum phenomena turn up in biology? Biology is, after all, a kind of applied chemistry, and chemistry is a kind of applied physics. So isn't everything, including us and other living creatures, just physics when you really get down to the fundamentals? (Al-Khalili & McFadden, 2014: 20).

Even here we can get stuck (productively!) with our exploration for perhaps a lifetime—our own. A computer, for example, certainly does not rely on any magic: it is explicable by physics. But is it derivable from physics?

And to which principles, simple ones at that, shall we appeal? After all, even science does not need them much for its daily activity. We quite agree with Bruno Latour (2017), who caricatures those who believe in the existence of universal, ahistorical foundations of Nature, those for whom facts are facts and Nature[3] obeys universal and eternal natural laws. He comments:

---

[1] The reader will find an attempt to explain some terms and expressions in the glossary at the end of the book. Often these will not be strict definitions, but rather the way they are used in the context of our book.

[2] Not natural history (*historia naturae*) that is not necessarily limited to physics. This distinguishes contemporary biology from earlier currents of description of the living that came not only from physics but were closer to philosophy or vitalism.

[3] Actually, this is not Nature and not even God: to please everyone, Latour uses an awkward acronym OWWAAB (Out-of-Which-We-All-Are-Born).

> And yet it is just as hard to reconcile this universality with the prodigious diversity of the scientific disciplines, the specialties, subspecialties, thematic networks and domains in which these 'unified' and 'universal' laws are applied in practice. [...] Viewed this way, the jungle of scientific descriptions looks more like the legal institution, with its complex casuistry of diverse codes and intermingled jurisprudences, than the unification implied in the traditional expression 'laws of nature'. [...] The process resembles the way legal precedents gradually take on importance through the multiplication of cases, trials, appeals, and counter-appeals, until the precedents invoked by the various courts of justice acquire the status of warranted, relatively universal principles—at least for as long as they are cited, archived, and interpreted (Latour, 2017: 162).

The other extreme—not necessarily the opposite—declares the whole cosmos to be alive, from elementary particles to the universe, and in this hylozoic continuum life is then only a kind of sub-region. We can refer, for example, to Renaissance doctrines of sympathy, pointing out that every entity "knows" at every moment what is happening in the whole universe, and similarly we can consider even artificial intelligence a living being. Let us not treat life so broadly: let the reader understand that throughout this text "living" an "life" refer to those manifestations of reality that are based on the existence of cells, i.e., are part of the biosphere(s) that appeared here on the planet about 4 billion years ago. Whatever that falls today under the rubric "life" is a descendant of that ancient gush; all living beings are therefore related to each other. It follows that neither molecules (DNA, proteins) nor more complex non-cellular structures (referring particularly to viruses, which many consider to be alive) are alive, and of course neither are wind nor stars nor protons alive in this sense.

We still need to consider the fact that we ourselves are alive, and so we look at the living "from the inside", life does not simply lie in front of us as unbiased observers; this certainly complicates our perspective. But this fact can also be turned to our advantage if we try to extrapolate from our own experience—as living creatures—to all living things. Of course, this must not be exaggerated: one cannot extrapolate in this way from the apparent achievements, the evolutionary novelties (apomorphies) of the species *Homo sapiens*, especially the ability to think and speak. Even as we accept this, there remain any number of features that can be used to draw analogies from areas considered to be the explicit domains of the human way of life—and yet remain difficult to capture by standard biological methods. These mainly concern the ability to orient oneself in the world via the basis of memory and experience—which is to interpret the world (i.e., create representations of it). We are therefore suggesting that there are spheres of cognition of the living that cannot be subsumed under contemporary biology, i.e., under "biology" as the community of biologists generally conceives it. For if we start from humanity, we must as well consider what the human sciences esteem. This is primarily the interpretation of historical events in human societies (history) or in the biosphere.

What does biology and history have in common? Specifically, it is the investigation of past phenomena to which we have only a mediated and desperately incomplete access. This involves a great variety of observed and studied phenomena, as well as their striking variability—in space and especially in time. At the same time, it is true for both branches of research that people were able to notice this variability only after numerous methodologies and "memory aids" had been introduced in a certain way: as myth, books, laboratories, archives, scientific institutes, etc.

We will proceed on the assumption that a fundamental characteristic of all living things, yet hardly absorbable by natural science, is the ability to recognize and interpret signs. In the past, we have even attempted to define life as a semiotic category: "Life and semiosis are co-extensive" (Markoš, 2016; Markoš et al., 2009). Semiotics, the art or knowledge of interpreting the world, of recognizing and creating objects, is certainly not a *natural* science, yet it will be at the forefront of our interest. It draws on a long interpretative tradition describing human understanding—understanding speech, notations, signals, gestures, and other manifestations. Only in the last few decades have there been attempts to apply semiotics to forms of life other than humans, but these have mostly been limited to "higher" animals equipped with a central nervous system. If we have the ambition to extend this explanatory scheme—within the framework of biosemiotic investigations—to all living things, and if we want this text to appeal to historians *and* biologists, we must address the question of whether an analogy can be drawn from speech, as well as from writing and reading, to some other properties of the living. We will leave in the background other properties also applicable in this way, such as a concern for symmetries or even a sense of beauty.

The second main theme of this text, which builds on interpretation, is evolution, a historical process having lasted about 4 billion years. We will take inspiration from the historical sciences that study the history of societies and civilizations, and will try to grasp the evolution of life, which is mostly the domain of the natural sciences, from that perspective. We should distinguish, in this respect, between history (that which has happened to which we have no access) and historiography as an interpretation of what has happened, according to the recognition of preserved traces but also to the "momentary setting" of cultures and contexts—those of the past and especially today. *We usually know such contexts better than the former participants themselves, but also, unlike them, we know what followed after they passed away.* This applies to the history of human civilizations written by historians, as well as the history of life written by evolutionary biologists. There are in neither case universal laws from which to derive history, to decide definitively and forever, "how it really was", and to set out the trajectory of how the "system" will behave in the future; we are in the territory of forensic techniques. To make the text accessible to both "denominations" of scholars—historians and biologists—part of our explanation takes the form of examples and case studies.

Historians, after certain unsuccessful attempts to develop their field into a science equal to that of natural science (as in case of historicism or historical materialism), have learned that it is impossible to predict the future, or even to determine the laws of historic development. So too, evolutionary biologists certainly do not seek to discover future trends, but as far as the interpretation of the past is concerned, we suspect that many still believe that "one day" they will get a clear picture. After all, biology is a science, so one must either believe this or accept what other sciences have no problem with: history does not belong in (natural) science, so evolutionary biology necessarily bows out of it. Hence, the evolutionary biologist is in the situation of the historian, spinning up a story so as to help him/her better understand the messages of the sources. And if we acknowledge this, then it would be useful to look at how the historical sciences approach the topic. This endeavor has brought us, the authors, a historian (JH) and a biologist (AM), together to discuss life and its history, the ways in which historiographers and biologists examine their sources, and how, when confronted with current knowledge (and the state of the sources), they construct narratives of the courses of times past.

Doubts come immediately to the fore. Communities of the biological species *Homo sapiens* have, for as far back as memory goes, devoted themselves to interpreting their past: rather than waiting tens of thousands of years for the birth of historiography, they have approximated the past on the basis of memory and experience as well as their interpretation, in the form of narratives, myths, rituals religious as well as secular and sacral monuments, script and other practices. Based on this corpus, they determined their place in the world, and often anticipated developments for the immediate future. With the introduction of writing, this memory became more accurate, but really, who knows? For example, scripture is seemingly a reliable source, but at the same time the ambiguity of reading often allows multiple interpretations of texts, thus expanding the presentation of the past to a much greater breadth. Moving into the realm of evolution, very similar patterns can be observed there as well. On the one hand, there are universal structures and (genetic) notations common to large groups of living things, even the entire biosphere, and on the other, there is a creative elicitation of what can be done with this equipment in the world into which a given living being was born.

And do we really want to take our analogies seriously, and apply what we know of humans to all living things? If so, then in addition to historians and evolutionary biologists there is a third group of interpreters of history. We will ask in all seriousness whether similar interpretative schemes play a role in non-human species—for example, in the parasol mushroom (*Macrolepiota procera*) or in the bacterium *Yersinia pestis*. We ask the same about communities comprised of collections of species (including humans), such as populations and ecosystems (e.g., a holobiont or a meadow). In this text we will keep in the background the assumption (hypothesis) that this might be the case. This approach is not in the least intended to deny other biological interpretations—after all, one of us is a biologist and bows to the results of all branches of biology, including evolutionary biology. We are merely looking for a niche from which all these achievements can be glimpsed in another way than through the views prevalent in biology and history studies today—that is all. But this requires leaning out of biology and indulging in a view of life that includes areas of knowledge other than the (natural) scientific: hence semiotics and biosemiotics.

1.1 Theses

If we were to summarize the contents of the book in three points, they might be as follows:

1. Historians (historiographers) spin narratives. Unlike fiction writers, propagandists, etc., their narrative is kept within bounds through the veto of sources. What is a source, however, is determined by community agreement.
2. Evolutionary biologists spin narratives. Unlike fiction writers, propagandists, etc., their narrative is kept within bounds through the veto of sources. What is a source, however, is determined by community agreement.
3. This rule is also a genuine subject for the field of biosemiotics; it can be applied to all living things.

1.2 The Model of Two Closures

We seek a common platform for the historical events concerning all life, and our starting axiom will be that the history of human civilizations is itself a subsection of the evolution of life on the planet; in other words, we are a part of the biosphere (hence, subjects of biosemiotics, too). The task of our text is to point to this unity. This axiom is necessarily linked to our second axiom, that *all life is endowed with an innerness and a historical experience that allows it to interpret the world and its place in it*. The ability to interpret the world is therefore a fundamental prerequisite of life, whereas non-life definitely has no such ability. We therefore introduce the axiom that life is necessarily separated from the rest of the world by two kinds of boundaries—we call them Closures, and both Closures are inseparable from life from its very beginning (Švorcová & Markoš, 2023):

Closure 1 is represented by the interior of the cell. It constitutes an unconditional prerequisite of life; a disruption of the membrane separating the interior from the environment means the death of the cell. This membrane acts as a barrier securing the internal structure and composition of the cellular environment; it acts as a selective filter for the entry of substances and most types of energy; it has sensors informing the inner space about the state of the external environment as well as means by which it can broadcast that information to the community of other cells (e.g., Baluška et al., 2022). In addition to the conservative, yet dynamic and time-varying structures, the cell has "invented" a way of writing information about these structures in the form of a linear notation made by nucleic acids; this notation, we emphasize, is *not* identical to the structure in question; moreover, it can be copied relatively easily and with high fidelity. Retroactively decoded (with a dose of interpretation according to the contexts in which the cell is found) this "text" provides the cell interior with instructions how to produce cellular and multicellular forms.

Closure 1 has never ceased to exist since the origin of life, and this is true for all contemporary life forms. There is a continuity of cellular life on the planet and of the associated compliance with certain rules and norms valid for all historically formed life lineages. The norms may have differentiated in each of such lineages into many "dialects", with any number of "additions", whether in the form of new interpretations or new structures and apomorphies (Markoš & Švorcová, 2019, for details). Importantly, such established variants of norms are usually and at least to some extent intelligible to cells in all lineages. This is, of course, a prerequisite for the existence of the second type of Closure.

Closure 2 is represented by various forms of multicellular communities (assemblages), which are established historically and exist for different durations, but are also established *ad hoc*: their interior is thus based on different principles than the "immortal" Closure 1, yet is always and necessarily linked to it. Closure 2 has a memory based on consensus and negotiation by its components (cells or multicellular assemblages), and as we shall see, it can always and in every respect be described with the label "culture". It is often similar to a notation, which preserves the memory and experience of a culture, but is not found only in the speech and texts of human cultures: memory traces and their interpretations abound in other lines of life—namely ethological, ecological, or ontogenetic rules and practices—which is what opens up the possibility to claim that Closure 2 is actually synonymous with culture. For obvious reasons we do not want to provoke users of other meanings of the word, hence we coin a new, neutral term "Closure 2".

To many anthropologists, historians, or sociologists, such a broadening of the meaning of "culture" may seem unacceptable. Rather, culture should be understood as *our* theoretical concept by which *we* interpret certain manifestations of members of a particular community. For we can never know for certain that the community we are studying (as a whole) has a particular culture, but to interpret the manifestations of a community in an appropriate way, we must simply assume that it developed one. In this way, it's *we* who bring into the field of research the assumption of at least a minimum degree of cohesion within the community studied, a minimum taking place within it that is both necessary for semiotic processes and propagated by social learning. While we understand this methodological caution, we nevertheless think of culture in an ontological sense (as being possessed by a given community) because and insofar as we accept Deely's proposition that semiotic relations are real. We therefore ask readers to share this (work-simplifying) approach with us in the pages of this book. Let us now return to Closure 2.

In nearly every case, a living entity can simultaneously belong to a number of quite disparate Closures 2, and belong to them with a varying degree of strength and importance: this network of Closures constitutes the basis for the functioning of the biosphere. Closure 2 thus consists of cell communities that come and go: a bacterial coating, a multicellular organism, an ecosystem, a family, a nation, or a sports fan club, to name a few. The norms and their interpretations in these Closures may vary according to contexts, and their stability and durability may also vary. This also means that an existing Closure of one kind can become part of Closures of other kinds (quasi-persons—see 5.5). For example, the multicellular Closure 2—a man as an army officer—is a part of the army for his entire professional life; at the same time he has a family in which his military rank is *quite* secondary; his whole body is an ecosystem containing both the multicellular (human) organism with its own ontogeny and the multispecies community (ecosystem) of the holobiont (wherein the rank and family play a *highly* mediated, but not insignificant, role); he also considers himself an active member of various societies and lodges. These various Closures 2, however, are not grouped hierarchically; they influence each other in a complex web of historical, ecological, educational, nutritional, familial, and other relationships.

There is no doubt that the two Closures communicate with each other in a variety of ways, and the flow of information between them is not unidirectional, as is often dryly assumed. Certainly, it is the cells that underpin every Closure 2, but information also flows backwards, in complicated feedback networks of various duration. Closure 2 communities can also influence the inheritance of cell lines: this knowledge is recent, and it is hidden under rather fuzzy terms such as the "inheritance of epigenetic settings", or "somatization of culture". We will mention such processes at appropriate places.

Why this division into Closures? Scholars of various disciplines—of both natural sciences and humanities—may have difficulty accepting our argument. Why, for example, is the "basis" of a community not a human being? Then "Closure 2" could represent various societies, nations, linguistic communities, etc. Why stuff these noble societies into the same bag with, say, bacterial coatings, plants, or fungal mycelium? We would like to persuade the reader that all such examples (and many more) fall under the common denominator *cultures*, and therefore are endowed with *biosemiotic* abilities to interpret their world. Such communities are endowed with a collective participative memory disseminated by social learning.

We urge the reader to accept our two-domain model and try to see by this optics some characteristics of life in general in this way. One of the reasons for the division is the endless and mostly futile verbal sparring, to give just one example: does a plant perceive or just feel? Or—in case of animal perceiving: is it confined to animals equipped with brains, or even more strictly to vertebrates? Why are there conferences on plant neurobiology, or perhaps even plant psychology, where the participants spend most of their time in sterile debates aimed at defining of utmost basal concepts? Or from the other end: cultural anthropologists do not like to be told that they are in fact ecologists. And ecologists would also raise their eyebrows at such a statement; they hardly consider, for example, a forest to be a culture. And is it that the thinking of our venerable philosophers is likewise rooted in the cultural behavior of their gut microflora? Even in a language environment we often cannot come to an agreement, let alone pair very similar concepts between different languages. The main difficulty is that these are often represented by international words with an extraordinarily wide field of meanings. Moreover, there is the second meaning of the word "culture", derived from agricultural cultivars and model laboratory organisms. The word "Closure" is not, we believe, as loaded as "culture", so we decided to make this cut between cellular and supracellular manifestations of life in this way, and use this awkward but to our knowledge unused terminology—hence "Closure 1" and "Closure 2". We considered, for example, using the doublet cell-ecosystem instead of the cumbersome Closure numbering, but again we face misunderstanding: for example, is a "nation" an ecosystem? And would these be mere analogies? Clearly the discomfort would be the same as saying "Closure 2".[4]

Closure 2 is never a pile of mutually indifferent cellular or supracellular entities; as we dare to argue here and elsewhere (especially in Chap. 10), it represents a culture. The affinity of one individual to different Closures refers to the aforementioned network of cultural relations, where each of these clusters places a different emphasis on certain aspects of being together in the world; in each of the Closures, the individual participates in viewing and constructing the world in a different way—and they coin the character of its form. Thus, the entire biosphere, including human civilizations as we understand them (and for the purposes of this study), is a network of Closures 2, not a hierarchy of nested entities as it is usually portrayed. In comparison, the previously mentioned random assemblages of living creatures (say, a poultry house or a crowded tram),[5] as well as the physical world, represent "non-cultures": in relation to such as these, culture can be defined.

The important point is that the two—interrelated—types of Closures have existed *from the very beginning of life*. For both types of Closures, the primary issue is the manifestation of internal dynamics and structuring as they have developed over time; on the basis of which they are modified (or modifiable) by external circumstances, which are substantially determined by other Closures to which the subjects of the Closure under study may or may not belong. Thanks to the Closures, each form of life "takes care" of its own affairs, it is not passively dragged along by life; it negotiates its place in the world—which is to say it understands the world, at least to some extent. The following chapters talk about the tools that life-forms use to do this, and offers examples of cooperation between Closures—undoubtedly a rich subject matter for biosemiotics, as is for its sub-areaof culturology. We will also mention the tools we use to study all this.

Thus, we are looking for ideas—analogies—to help us characterize all living things and *nothing but* living things. We have discussed several familiar metaphors deeply embedded in our culture, but none is strictly selective. Metaphors such as "part versus whole", "individual versus community", "nature versus culture", "higher and lower", etc., are suited not only to living creatures but nearly

[4] The notion of Closure was taken from the archaic meaning of the term and introduced to these conversations by Stuart Kauffman (2000) to separate the interior of an "autonomous agent" from its environment, and the notion has since taken hold. Kauffman did so rather glibly; the context of his book perhaps mostly implies that the "autonomous agent" is a cell. Nevertheless, numerous examples point to the "bargaining" of autonomous agents regarding the progress of the "biosphere" from the "actual" to the "adjacent possible"; he also gives examples of such negotiations between ecosystems and between human societies (economics).

[5] The empress Maria Theresa undoubtedly represented a multiple of Closures 2 (as a community of cells descended from a zygote, as the ecosystem of a holobiont, as a member of the Catholic Church, as the mother of a family; so too was the Austrian Empire she ruled). We try to avoid hierarchizing such Closures 2; the only condition for their existence is that they presuppose the constituted organized and interacting Closures 1. Thus, Closure 2 is not a cube of yeast from the supermarket, albeit full of cells, nor is it a pile of freshly cut grass (also full of Closures 1 and 2); rather, Closure 2 is a sourdough and a meadow.

everything, and could thus dilute, if not shatter, our approach. When such nested levels are consistently tightened, we could easily land at the view that even the whole of human society will one day be explicable from the level of atoms, or even the subatomic world, because it stems from the scientific logic.

Digression one, which alone has the power to challenge such an optimistic model. Real systems are usually open, exchanging with their surroundings not only energy but also matter; moreover, they usually occur in steep gradients of both, so that disturbances of homogeneus state tend to amplify and form macroscopic dynamical (so-called dissipative) structures, as flames, stars, or hurricanes. Suddenly the history of such a system starts to play a role here: its evolution requires reconstruction. A model of life? Many physicists and physical chemists, including the author of the theory of dissipative systems, I. Prigogine (e.g., Prigogine & Stengers, 1977, 1985), have flirted with this possibility.

A second digression concerns the informational side of living systems. Inanimate structures (dissipative structures as, e.g., river meanders) have no memory of their past other than their structure itself. The genetic notation and multilevel organization of structures, processors, and "information technologies" necessary for the operation of cells are missing; they are not needed for the function of these structures. And the ability to interpret contexts and situations on the basis of memory and experience does not at all fit into a purely scientific description of molecular genetic processes.[6] Life therefor remains a mystery to science, and we believe it cannot be explained by existing scientific models—perhaps precisely because, as we have shown, history does not belong to science. This is also why we have attempted to create an alternative model of the two Closures that will accompany us in the pages of this book.

Again, why Closures? Cells do not arise de novo, nor have they always been there: cells are *born* only from other cells and are themselves capable of giving rise to new cells before they die—and this is how it has been since the beginning of life. The continuous existence and yet *uniqueness of every single cell* establishes Closure 1 in relation to the external, inanimate—but also and above all, the living world. A community of cells is "homogeneous" only in very rare cases, for example in well-treated laboratory models (see especially 8.4). To repeat: cells enter into elaborate structures and interactions with other cells to form communities—Closures 2 of different types and durations; the same cells can be part of different types of Closure 2, as we discussed above. All these Closures 2 are formed by the activity of cells and cell communities. They can persist, often for very long periods of time, but they do not have the "eternal duration" of the cell lineages of Closure 1; they appear and disappear, and similar ones can re-emerge and persist for some time. The reader will

[6]We don't want to resort to silly clichés such as this: "In the 1920s, life was still a mystery. Although nineteenth-century biochemists had made great advances in constructing a mechanistic understanding of the chemistry of life, many scientists continued to cling to the vitalist principle that biology could not be reduced to chemistry and physics but required its own set of laws. The 'protoplasm' inside living cells was still considered a mysterious form of matter animated by unknown forces" (Al-Khalili & McFadden, 2014: 50).

note that Closure 2 is diametrically opposed to physical systems such as dissipative structures. So let us try to accept the suggestion presented above that Closures 2 represents culture—and thereby complete our model. Though of course, like any model, it also represents a reduction of the world.

Let us therefore put all these forms of cellular communities on a common platform and ask what they have in common—as living beings. They may represent (highly) organized units, which implies symbiosis—a high degree of coordination among the cell-members of the community (and the exclusion of non-members). But there can also be short-term communities based on adherents of fashion, fans of anything, or ad hoc alliances established in response to this or that situation. As already said above, a single organism is also part of several different communities (for example, the multicellular organisms of humans are part of a language community, and also part of the ecosystem of the holobiont—each of which are different forms of Closure 2).

Thus, Closure 2 implies a considerable degree of understanding between members of a given community, across several domains simultaneously. Most of all, the various lineages of living things across and as well as the whole biosphere are descendants of ancient cells from the primordial biosphere: since life emerged on the planet, they exist not only in Closure 1 but also, as we have already mentioned, in various forms of Closure 2, which are constantly appearing and persisting for various lengths of time. Both Closure levels carry memory (genetic and epigenetic) and experience (of the individual, of the lineage they belong to, plus the imprinting of the environment): let us summarize this under the common name: the space of experience.

Figure 1.1 approximates the idea of Closure 2 with an analogy to the game Mikado. Each stick in the heap represents a different Closure 2, and the whole heap is then a set of such Closures 2 of which a cell, as a necessary part of any community of Closure 2, is simultaneously a member. Pulling or adding sticks changes the living thing's picture of the "world", and the position of the stick is a measure of its importance. Those quite at the top are easy to add or remove (for example,

Fig. 1.1 Mikado as an analogy of the set of Closures 2

belonging to a sports club, to a school class, or to a set of supporters or voters of anyone or anything). The lower the sticks are placed, the harder it is to add or remove them, and the more the whole edifice threatens to collapse when they are mishandled. At the very bottom, for example, is the organization and early ontogeny of the multicellular body—pulling out such a stick may even cause the death of the individual (for example, in the case of certain malformations, tumors, or organ failures); and a new one cannot be obtained. Closures 2 somewhat higher in the pile may turn out to be interchangeable, but only with difficulty: family, language, life in the sea. Or a lineage may produce new "sticks" leading to new species. Still higher is occupation, employment, wedding ritual, etc. An individual can easily change his/her affinity for this or that Closure "near the top" according to the situation: a die-hard Arsenal fan forgets his family and occupation during a thrilling match, for example, but again, if Arsenal were disbanded, family and occupation would usually not be shaken overly much. Yet all the while our fan remains a multicellular Closure 2—an individual of the species *Homo sapiens*, influenced by the other Closures of which he is a member (see also Chap. 3). The game never stops, sticks are drawn out and others added, some are rearranged and may even sink to lower layers.

The synchronously living members of *each* Closure 2 community claim their space of experience and thus create a unique, commonly held horizon of expectations. To which point at the horizon individuals and communities will move (and along what paths) is a matter of negotiation and convention—and only life exhibits this capacity for negotiation! Of course, during the progress towards the horizon, various breaks can also occur (catastrophes, modification of the horizon…). The breaks can be caused by unforeseen external factors (epidemics, climate change, etc.), or they may be produced from within. Such singularities, of course, always carry with them an element of uncertainty (often considerable) and can cause anything from the demise of a community to the emergence of entirely new life forms.

Cells, lineages and communities are subject to evolution (mostly divergent; Markoš, 2016), and in their space of experience they maintain various tools of communication (i.e., information transfer) common to the whole biosphere, or to a certain lineage or type of community. They have the ability to process and interpret the transmitted information—if they consider it relevant; this includes the possibility to interpret it incorrectly, but also correctly yet in a completely novel way. Thus, there is always a certain degree of overlap in the experience of all the players in the community that allows for a game of bargaining (Markoš & Švorcová, 2024). In this context, we will distinguish between signal, sign and relation. Information requiring no interpretation is a signal; the recipient's response to the signal is either unambiguous or, depending on the context, representing one of several pre-known possibilities (distributed statistically in any given setting). Information that requires interpretation is a sign; we will argue again and again that the ability to recognize signs and interpret them is a defining feature of life. And into the final interpretation of the sign also enters the relation between the internal setting of the two Closures and the sign itself: it is at this point that we straddle the boundary between what natural science describes and the domain of social sciences (i.e., the sciences concerned with human societies).

Box
In recent years the trio of authors F. Baluška, A. Reber and W. Miller have presented a number of articles and monographs that have emphasized the interior of the cell and its unbroken continuity since the origin of life (e.g., Baluška et al., 2022; Reber et al., 2023; Reber & Baluška, 2021), which is seemingly close to our perspective. They consider the plasma membrane as the central structure (i.e., in our view the mere "organ of Closure 1"), which has not ceased to exist throughout the evolution of life. But this is where the similarity with our view ends. If we understand the authors' argument well, they assume something outside the cell as an objective reality (is it the world itself, or a model of it?) knowable "in reality" as it is. The plasma membrane is then a filter that picks up some properties of the world and transposes them in the cell into a "body language", de facto adjusting internal processes according to the context, according to the semantics of the incoming information. Moreover, something of this interpretative process can also be stored in the genome and made available to future generations; here there is a loose analogy with our Closure 1. But as with the the external world, the set of intracellular processes—the *senome* in their terminology—is described in purely mechanistic terms, allowing hypotheses to be formed and tested. The second part of their model is intercellular communication, the "N-dimensional episome"; this guarantees the behavior of multicellular formations—generated by cells and simultaneously controlling their further behavior. Again, perhaps very loosely Closure 2, without the further subdivisions of the kind we depict above. The evolution and semantics of life processes focuses on the senome (or episenome) i.e., the membrane and membrane continua, the rest is—physics. M. Barbieri's *organic codes* (2003) may look very similar when he claims that "codes are not metaphorical, they are real". Unlike Barbieri, however, the authors mentioned above are convinced that a mechanistic model will be able to show that "mental faculties" appeared in the very first cells and, thanks to this, are found in all inhabitants of the biosphere and are not an invention of our species. Words such as "sensing", "perceiving", perhaps "sentience", even "mind" and the like, seem to represent synonyms or a continuum of very blurry meanings (a glance into any English explanatory dictionary even encourages that they are not terms, but words literally oversaturated with many meanings).

1.3 Basics to Be Taken by the Reader

1. All life is based on cells. Life, and only life, is a semiotic category, i. e. all semiosis is associated with life (we do not consider hylozoic, panpsychic, physiosemiotic etc., views as semiotic). Therefore, we take "semiosis" and "biosemiosis" essentially as synonymous: *all* semiotics is necessarily biosemiotics.

2. (Bio)semiotics is based on interpretation. It follows that historiography and evolutionary biology are analogous, interpretative disciplines, i.e. they reconstruct history of life on the Earth. This holds in spite of differences in their methods, terminology, and history of both branches of human knowledge. The biosemiotics proper is based in the statement that *all* appearances of life are able to generate and interpret signs, hence understand their “being in the world”. It follows that mind and articulate speech are not necessary conditions for semiosis. Such a condition is only single: to be alive.
3. Above statements suffer a serious drawback that, understandably, all semiotic (hermeneutic, forensic, diagnostic, etc.) models were developed by humans, and for usage under human conditions (just consider the ways of usage of the couple “thing/object”). It was not our goal to go into comparisons of many different models existing in semiotics, hermeneutics, and philosophy. Instead, we decided to focus on a single explicatory model as developed by John Deely, who is the follower of the venerable tradition of the Latin civilization, punctuated with names as Augustin, Aquinas, Poinsot and/or Peirce. This allowed us to devote our attention to main topics of the book—historiography and evolutionary biology: in our opinion, they fit into sub-disciplines of biosemiotics.

References

Al-Khalili, J., & McFadden, J. (2014). *Life on the edge. The coming of age of quantum biology*. Bantam Press.

Baluška, F., et al. (2022). Cellular sentience as the primary source of biological order and evolution. *BioSystems, 218*, 104694.

Barbieri, M. (2003). *The organic codes. An introduction to semantic biology*. Cambridge University Press.

Ginzburg, C., & Davin, A. (1980) Morelli, Freud and Sherlock Holmes: Clues and Scientific Method. *History Workshop, 9* (*Spring*), 5–36.

Kauffman, S. A. (2000). *Investigations*. Oxford University Press.

Latour, B. (2017). *Facing Gaia. Eight lectures on the new climatic regime*. Polity Press.

Markoš, A. (2016). The birth and life of species–cultures. *Biosemiotics, 9*, 73–84.

Markoš, A., & Švorcová, J. (2019). *Epigenetic processes and the evolution of life*. CRC Press.

Markoš, A., & Švorcová, J. (2024). Of bubbles and foams: Umwelt counterpoints in symbiosis. *Sign System Studies, 52*, 438–456.

Markoš, A., et al. (2009). *Life as its own designer. Darwin's Origin and Western thought*. Springer.

Prigogine, I., & Stengers, I. (1977). The new alliance. *Scientia, 112*, 259–302 and 643–652.

Prigogine, I., & Stengers, I. (1985). *Order out of chaos*. Flamingo.

Reber, A. S., & Baluška, F. (2021). Cognition in some surprising places. *Biochemical and Biophysical Research Communication, 564*, 150–157.

Reber, A. S., Baluška, F., & Miller, W. M. (2023). *The Sentient Cell: The Cellular Foundations of Consciousness*. Oxford University Press.

Švorcová, J., & Markoš, A. (2023). Closures as a Precondition of Life, Agency, and Semiosis. *Biosemiotics, 16*, 45–59.

Chapter 2
Charade from One Pocket

Abstract A short detective story by K. Čapek as an appetizer will shoe main topics of the concern of this chapter: recognizing incomplete and scattered data, interpreting them, and constructing narratives from. Considering, one, what the interpreter knows before starting his/her research, plus his/her presuppositions and biases; second, what intentions comply him to analyze them; and third, how to put the data into context with contemporary needs or demands of his/her community. Finally, an inventive ("forensic") aspect and contingency also play a role. All this must be confronted with the veto of sources. Figuration, canonic texts, and hypolepsis as tools of world–making, but also of mutating ("rewriting") it. Structures, codes, and interpretation in the evolution of life. Six theses of narrative interpretation of human history, valid perhaps also to all other forms of life. Underdetermination of theories. Interpretation is not translation. Historism and narrativism, language and the world—demarcation line between them.

Keywords What is data · Intentions and presuppositions · Narrative · Contexts of *small world* · Facts, clues and methodology · Hypolepsis

> The fact that it does no good to speculate about what might have happened in bygone ages is underlined by the fact that such speculation always takes the form of a narrative. If the aim of science is to see how much one can explain by rules governing observable phenom ena, however, a scientist must avoid giving up the search prematurely simply by postulating ad hoc, and question begging, stories that could have accounted for the occurrence of something long ago. Science is not narrative. As in physics, the rules or laws governing changes in the earth must be presumed to be the same throughout time. What goes on in the world, therefore, is an endless series of law-governed instances rather than a concatenation of episodes.
>
> David J. Depew & Bruce H. Weber, 1996: 97.

A. Markoš, J. Horský, *The Seeds of Forms and the Forms of Seeds*, Biosemiotics 12, https://doi.org/10.1007/978-3-032-16131-4_2

As an appetizer, we start with a short story by Karel Čapek that demonstrates some problems (perhaps the main problem) of historical research. The next chapter will discuss the role of interpretation in history studies.

2.1 Karel Čapek: The Fall of the House of Voticky

Karel Čapek K (1994: 103–113)

One day there arrived at the office of Detective Captain Dr. Mejzlik a smallish man with gold-rimmed glasses and a worried expression. "Archivist Divisek," he murmured "Captain, I've come to consult you… as a prominent criminologist; that is, I've been told that you happen to be… that is to say… most particularly in regard to the more complex cases—That is to say, this is an unusually puzzling case," he declared emphatically.

"Just tell me what it's all about," said Dr. Mejzlik, reaching for a notepad and pencil.

"There ought to be an investigation," archivist Divisek blurted out, "as to who murdered Sir Petr Berkovec, how his brother Henricus died, and what happened to Sir Petr's wife, Katerina."

"Berkovec, Petr," Dr. Mejzlik reflected "To my knowledge we've received no report of his death. You'd like to make a statement, is that it?"

"Oh, no," the archivist said. "That is to say, I've only come to you for advice, you see. Something terrible must have happened".

"When was it?" Dr. Mejzlik prompted "First of all the date, please."

"In the year fourteen hundred and sixty-five, of course," said Mr. Divisek, looking reproachfully at the captain through his glasses. "You ought to know that, sir: during the blessed reign of King Jiri of Podebrady."

"I see," said Dr. Mejzlik, and he tossed the notepad and pencil aside. "In that case, my dear sir," he stated with conspicuous cordiality, "it would be better to talk with Dr. Knobloch," by which he meant the police physician. "Why don't I call him?"

The archivist frowned "What a shame," he said, "since you were so highly recommended to me. You see, I am in the midst of preparing a historical work on the reign of Jiri of Podebrady, and when I stumbled, yes, stumbled across this matter, I didn't know where to turn for advice."

Harmless, Dr. Mejzlik decided "My dear sir," he said quickly, "I'm afraid I can't be of much use to you. I must confess that I'm far too weak in history."

"That's a mistake," Mr. Divisek noted sternly. "You ought to know history. But since the relevant historical material is not known to you at first hand, sir, I will introduce you to all the known facts; unfortunately, there are very few of them. First and foremost, there is the letter from Sir Ladislav Pchac of Olesna to Sir Jan Borsovsky of Cercany. You are, of course, aware of this letter."

"I'm sorry to say that I'm not," Dr. Mejzlik confessed with the contrition of an ill-prepared schoolboy.

"But, my good man," Mr. Divisek burst out in astonishment, "the historian Sebek published this letter 17 years ago in his *Regesta*! At the very least you ought to know *that*! However," he added, adjusting his glasses, "neither Sebek nor Pekar nor Novotny nor anyone else devoted proper attention to this letter. And this very letter, which of course you ought to have known about, led me to the clues in this case."

"Aha," Dr. Mejzlik said. "Go on."

"First of all, the letter," said the archivist. "Unfortunately, I don't have the text with me, but only one of its references pertains to our case. He—that is, Sir Ladislav Pchac—writes to Sir Jan Borsovsky that his uncle—that is to say, Sir Jan's uncle, Sir Jesek Skalicky of Skalice, in this Year of Our Lord 1465, is no longer in attendance at the court in Prague, in view of the fact that His Royal Highness commanded Sir Jesek, 'due to the woeful deeds at Votice Velenova,' as the writer puts it, henceforth to come no more to the royal court, to repent of the Lord God his choleric temper, and to await divine justice. You understand by this," the archivist explained, "that we would say that His Royal Highness confined Sir Jesek to his own estates and holdings. Doesn't something about that strike you as odd, sir?"

"Not at the moment, no," Dr. Mejzlik said, drawing a remarkable spiral on his notepad.

"You see!" Mr. Divisek exclaimed triumphantly. "Sebek didn't notice it, either. It is a *very* striking circumstance, sir, that His Royal Highness did not summon Sir Jesek—whatever those woeful deeds might have been—to proper secular judgment, but committed him to divine justice instead. His Highness makes it unmistakably clear," the archivist continued, with obvious deference, "that these woeful deeds are of such a nature that the sovereign himself exempts them from worldly justice. If you knew His Highness, sir, you would realize that this is an extraordinarily unusual occurrence. King Jiri of blessed memory was most firm in his regard for the strict and proper discharge of justice."

"Perhaps the king was afraid of Sir Jesek," Dr. Mejzlik offered "After all, in those days—".

The archivist sprang from his chair in indignation. "Sir," he stammered, "what are you saying? That King Jiri might have been afraid of someone? And of a mere knight at that?"

"Then there was some sort of favoritism involved," Dr. Mejzlik said. "You know, the way we—".

"No favoritism!" Mr. Divisek cried out, quite red in the face. "Perhaps one could speak of favoritism during the reign of King Vladislav, but during the reign of King Jiri—no indeed, sir, how could you even think of such a thing! He would have banished you for even suggesting it!" The archivist calmed himself somewhat. "No favoritism. Sir, it's obvious that there must have been something singular about those woeful deeds for His Royal Highness to have abandoned Sir Jesek to divine justice."

"And what were those deeds?" Dr. Mejzlik sighed.

Archivist Divisek looked surprised "But that is what you have to find out for me," he said in amazement. "What is a criminologist for? That's the reason I've come to you!"

"Oh, for heaven's sake," Dr. Mejzlik began defensively, but the archivist would not allow him to finish. "First you must know the facts," he instructed "When I noticed those obscure references in the letter, I began to investigate those woeful deeds at Votice Velenova. Unfortunately, no records have survived; however, in the church at Votice Velenova I found the tombstone of Sir Petr Berkovec, and that stone, sir, is dated 1465! As you know, Sir Petr was the son-in-law of Sir Jesek; that is, Sir Petr married Sir Jesek's daughter, Katerina. Here is a photograph of the tombstone—now, don't you notice something peculiar about it?"

"No," Dr. Mejzlik said, examining both front and back of the photograph of the tombstone, on which was carved a knight with his hands crossed on his breast, surrounded by an inscription in gothic letters. "Wait, there's a fingerprint here in the corner."

"It's probably from my finger," the archivist said, "but notice the inscription!"

"Anno Domini MCCCCLXV," Dr. Mejzlik read with some effort. "The Year of Our Lord 1465. That's the year of this man's death, right?"

"Of course it is; don't you know anything? But take a look: some of the letters are noticeably larger than the others!" And the archivist quickly wrote with the pencil: ANN**O** DOMINI M**CC**CCLXV. "The sculptor deliberately made the letters O, C, and C larger; it's a cryptogram, do you see? Write down the letters O, C, C—doesn't anything strike you about them?"

"OCC, OCC," Dr. Mejzlik muttered, "that could be—aha, it's short for OCCISUS, right? That means 'murdered'!"

"Yes," the archivist proclaimed solemnly. "That is the manner in which the sculptor indicated to posterity that the nobly-born Sir Petr Berkovec de Wotice Welenova had been murdered There we have it!"

"And his father-in-law, this Jesek Skalicky, murdered him," Dr. Mejzlik declared in an unexpected burst of historical enlightenment.

"Nonsense," Mr. Divisek said scornfully. "If Sir Jesek had murdered Sir Petr, His Royal Highness would have summoned him before the king's bench. But that's not all, sir: directly beside this tombstone is another, under which lies Henricus Berkovec de Wotice Welenova, the brother of Sir Petr; and on that stone is the very same date, 1465, only without a cryptogram! And on that same stone, Sir Henricus is shown with a sword in his hand; the sculptor obviously wanted to indicate that Sir Henricus had died in honorable battle. And now tell me, for heaven's sake, how these two deaths are connected!"

"Perhaps it's only a coincidence," Dr. Mejzlik suggested uncertainly, "that this Henricus lost his life that same year—".

"Coincidence!" the archivist shouted in exasperation. "Sir, we historians do not acknowledge coincidence! Where would we be if we admitted that something had happened by coincidence? There must be some kind of causal relationship! But I've not yet finished telling you everything. In the following year, 1466, Sir Jesek Skalicky died; and, if you please, his holdings at Skalice and Hradek fell to his cousin, Sir Jan Borsovsky of Cercany, whom I mentioned before. Do you realize what that means? It means, sir, that Sir Jesek's daughter, Katerina, who as every child knows was married in 1464 to this same Sir Petr Berkovec, was no longer living! And the Lady Katerina, if you please, has no tombstone anywhere! Would you say it is also coincidence that the Lady Katerina utterly vanishes from our sight immediately following the death of her husband? What about that? Do you call that coincidence, sir? And why is there no tombstone for her? Coincidence? Or does it have to do with those woeful deeds for which His Royal Highness committed Sir Jesek to divine justice?"

"It's altogether possible," Dr. Mejzlik supposed, with somewhat greater interest.

"It's altogether certain," Mr. Divisek proclaimed, admitting no doubts. "And now, you understand, it is a question of who killed whom and how everything is connected The death of Sir Jesek does not interest us, because he survived those 'woeful deeds;' otherwise, King Jiri would not have bid him repent of them to the Lord God. We must establish who killed Sir Petr, how Sir Henricus died, how to account for the Lady Katerina, and what Sir Jesek Skalicky had to do with all this!"

"Wait," Dr. Mejzlik said, "let's draw up a list of the parties involved:

(1) Petr Berkovec—murdered; (2) Henricus Berkovec—killed in battle, right? (3) Kate—vanished without a clue; (4) Jesek Skalicky—abandoned to divine justice. Is that it?"

"It is," said the archivist, blinking cautiously. "You might have said Sir Petr, Sir Jesek, and so forth. But go on."

"At the same time," Dr. Mejzlik reflected, "you eliminate the possibility that this Jesek could have murdered his son-in-law Petr Berkovec, because in that case he would have been brought before a jury."

"Summoned before the king's bench," the archivist amended "Otherwise, that is correct."

"Then that leaves only Henricus, Petr's brother. Most likely Henricus murdered his brother—".

"That's not possible", grumbled the archivist. "If he had, they wouldn't have buried him in the church—at least not next to his brother."

"Aha. Then Henricus merely arranged for the murder of his brother and later himself was killed in some battle, right?"

"Why then would Sir Jesek have received a reprimand from the king for his violent temper?" retorted the archivist, fidgeting in dissatisfaction. "And where would Katerina have disappeared to?"

"That's true," Dr. Mejzlik muttered "Listen, this is a complicated case. Let's suppose that Petr caught his wife Katerina *in flagrante* with Henricus and killed her. Her father, Jesek, found out about it and murdered his son-in-law in anger."

"That wouldn't work, either," objected Mr. Divisek. "If Sir Petr had killed Katerina for adultery, her father would have agreed with what he'd done. Believe me, in those days they were quite strict about that sort of thing."

"Wait a minute," said Dr. Mejzlik, deep in thought. "Let's say that Petr simply killed her; maybe he got into an argument with her—".

"But then they'd have erected a tombstone for her," claimed the archivist, shaking his head. "That won't do. Sir, I have racked my brain over this for a year now, and nothing fits together."

"Hm," said Dr. Mejzlik, contemplating the rather short roster. "Confounded business. Maybe we're missing some fifth person."

"What would you do with a fifth?" asked Mr. Divisek reproachfully. "You don't even know what to do with the four we have!"

"Then it must have been one of those two who killed Petr Berkovec: either his father-in-law or his brother—I'll be damned!" Mejzlik shouted abruptly. "Listen, it's Katerina! She's the one who did it!"

"Oh, dear!" the archivist burst out in evident distress. "I didn't even want to think about that! Good Lord, could she really have done it? And what would have happened to her then?"

Dr. Mejzlik's ears grew red with concentration. "Just a minute," he said, and he leaped up, pacing the room in agitation. "Yes," he cried, "yes, I'm beginning to see it now! My God, this is quite a case! Yes, it's coming together now—and Sir Jesek has the leading role! Yes, the circle is closing! And that's why Jiri—Now I understand! Listen, he was one sharp operator, King Jiri!"

"He was," Mr. Divisek stated piously. "Let me tell you, he was a wise monarch!"

"Look," said Dr. Mejzlik, sitting down on his desktop and landing on the inkstand, "this is how it could have happened; and I'd swear on a stack of Bibles that it's exactly what did happen! Now first of all, any hypothesis we advance must take into account all of the known facts; nothing, not even the slightest circumstance, can be allowed to contradict it. In the second place, these facts must be arranged into a single, continuous action; the simpler, the more contained and connected the action, the more likely that things happened exactly that way and not otherwise. We're talking now about reconstruction of the events, see. The hypothesis that places all known facts into the most connected, most rational chain of events is without qualification the one for us, do you understand?" said Dr. Mejzlik, looking severely at the archivist. "This is our basic rule of methodology."

"Yes," the archivist said obediently.

"Now the facts which we must bear in mind probably occurred in the following order: (1) Petr Berkovec married Katerina; (2) Petr was murdered; (3) Katerina disappeared without a tombstone; (4) Henricus died in some conflict; (5) The king rebuked Jesek Skalicky for his violent temper. (6) But the king did not have Jesek brought to trial: so Jesek Skalicky was somehow or other in the right.

"Are these all the known facts? They are. To continue: it follows from the known facts that neither Henricus nor Jesek murdered Petr. Who, then, could have murdered him? Katerina, obviously. This conjecture is supported by the fact that we have no tombstone for Katerina; probably she was simply buried somewhere like a dog. And why then was she not summoned for proper trial? Because, obviously, some hot-headed avenger killed her on the spot instead. Was it Henricus? Obviously not; if Henricus had punished Katerina by killing her, old Jesek might well have approved; but why then would the king have criticized Jesek later for his violent temper? It follows from this that Jesek, her enraged father, did away with Katerina. And now the question is, who killed Henricus in combat? Who did that, sir?"

"I don't know," the archivist sighed plaintively.

"Who else but Jesek?" shouted Dr. Mejzlik. "There's no one else who could have done it! That's the only way we come full circle in the whole affair, can't you see that? Look at it this way: Katerina, wife of Petr Berkovec, was—hm, how to put it—was burning with sinful desire for his younger brother, Henricus."

"Do you have evidence of that?" asked Mr. Divisek, enormously interested.

"It follows from a logical process," Dr. Mejzlik said decisively. "Listen, it's either money or a woman; you can take that as a given. To what extent Henricus returned that passion, I don't know; but we have to look for a motive here, a reason for Kate to have murdered her husband. And I can tell you right now," Dr. Mejzlik declared in an authoritative voice, "that she did it!"

"I suspected that," the archivist sighed gloomily.

"And now her father, Jesek Skalicky, arrives on the scene as avenger of the family honor. He kills his daughter because he doesn't want to turn her over to the hangman; and afterwards he challenges Henricus to a duel, because that unfortunate young man is more or less an accessory to the crime and to the damnation of his only daughter. Henricus falls in the duel, sword in hand. Of course there's an alternative here: Henricus shields Katerina from her enraged father with his own body and falls in the duel. But the first alternative is better. Thus you have the woeful deeds And King Jiri, because he perceives how seldom human judgment is called upon to judge an act so savagely righteous, quite reasonably abandons this appalling father, this violent avenger, to divine justice. A decent jury might have done the same. Within the year, old Jesek dies alone and grief-stricken, probably of heart failure."

"Amen," said archivist Divisek, clasping his hands. "That's how it was. The King Jiri I know could not have acted otherwise. Listen, that Jesek was a harsh and splendid character, wasn't he? It's all perfectly clear now; you can see it right before your eyes. And the way everything fits together," the archivist marveled "Sir, you have performed an invaluable service to historical scholarship. This casts such a dramatic light on the people of that era, and indeed—" Mr. Divisek, overcome with gratitude, threw out his hands with a flourish. "When my *History of the Reign of King Jiri of Podebrady* is published, I will take the liberty, sir, of sending you a copy, and there you will see how I deal with this case as a scholar."

Some time after that, Dr. Mejzlik really did receive the impressive volume titled *The History of the Reign of King Jiri of Podebrady*, with a warm inscription from the archivist Divisek. He read the entire book from A to Z, for—let us admit it—he was quite proud of his contribution to a scholarly work; but he found nothing there until, on page 471, among the bibliographical notes, he read the following: Sebek, Jaroslav, *Regesta XIVth and XVth Centuries*, p. 213, letter from Sir Ladislav Pchac of Olesna to Sir Jan Borsovsky of Cercany. Of interest is the comment concerning Sir Jesek Skalicky of Skalice, which merits particular attention but has not thus far been subjected to proper scholarly methodology.

2.2 Interpretation

Čapek's tale is a humorous historical "detective story", but it vividly illuminates what all narrative is about. Divisek, an archivist working on the history of the period of the Czech King Jiri (George, reigned 1458–1471), comes to the police to see Superintendent Mejzlik because he needs to clarify the somewhat indecipherable source data from that time—a brief comment in an old letter. The ancient writer naturally assumed that the events he was writing about (presumably a major crime) were well known to his contemporaries and therefore he needed not ruminate about them. He also thought that everyone *knowed* what he did (after all, it is so important!). So, as an *empirical author*,[1] he presupposes his *model reader* and compiles the report's data accordingly. Divisek (the *empirical reader*) cannot know these fine details after the passage of nearly half a millennium, although he certainly knows much more about that time than the chronicler (the *model author* for Divisek) could have known. Divisek also has the advantage of knowing the later interpretations of Jiri's reign—those that have piled up over five centuries—and from these he picks out the "correct" and "true" ones. However, he assumes that Mejzlik also knows just as much as *he* does (plus perhaps something extra, otherwise he wouldn't have come to him): "You ought to know history."

The starting point of the story therefore concerns knowledge of the context and, moreover, of the established (and perhaps, above all, newly established) evaluative clichés about the period in question. Even the historian who knows the sources perfectly is not immune to the schemas agreed upon by his own community. According to these clichés, King Jiri of Poděbrady was a "good" ruler, a fact that comes to the fore even more if the reader is familiar with the Czech history of the entire fifteenth century, in which the king and emperor Sigismund is the superior evildoer.[2]

Thus, from the perspective of our archivist, what prevails—despite his knowledge of the facts—is a priori piety and respect for a semi-divine ruler (a common motif in history of nations), who "cares" for his subjects and cannot even be imagined to be unjust let alone venal: "The King Jiri I know could not have acted otherwise." In this light, the monarch's lifelong intrigues, warfare, etc. are something not worth mentioning to Divisek. Rightly so, after all—*all* fiction is a so-called *small world*, which presupposes the knowledge and cooperation of the reader or listener (small worlds, see e.g., Eco, 1989).

There is another a priori here, this time from the small world of Detective Captain Dr. Mejzlik: "It's either money or a woman, you can take that as a given." Mejzlik knows his small-town Prague people and knows what to expect in his district: who

[1] See 4.4 for model/empirical author/reader.

[2] Understandably, this is a Czech choice of perspective: in terms of European history, Sigismund tends to be ranked as perhaps the most capable ruler of the Holy Roman Empire. For historians of other communities, the Czech history of the fifteenth century—including the Hussite Revolution, the Crusades and the turmoil following Sigismund's death—may appear as little more than a nuisance and deserving of only a brief mention.

knows how the conversation would have turned out if he had served in counterespionage or fought drug cartels.

The second perspective necessary for understanding the plot has a firmer foundation: the archivist knows the realities of the time; he can interpret what it means to be in "attendance at the court" or to receive a "summon before the king's bench"; he understands the cryptograms of the period (as OCC), courtesy phrases and iconography; he knows Latin; he is familiar with the laws and morals of the epoch (e.g., attitudes toward adultery); and he knows the available sources (letters, tombstones) necessary to construct the story. However, the archivist goes from "First you must know the facts" to another slip when he states that without sources, he doesn't make a move: "Where would we be if we admitted that something had happened by coincidence!"; or "Do you have evidence of that?"; even "As every child knows".

Mejzlik is thinking along similar lines when he speaks of the need to "take into account all of the known facts" and also "these facts must be arranged into a single, continuous action; the simpler, the more contained and connected the action, the more likely that things happened exactly that way and not otherwise." And let's also note the principle of parsimony, also known as Occam's razor: we have four players, and our task is to solve the puzzle with just these four: "What would you do with a fifth?", or "We're talking now about reconstruction of the events, see." And then they proceed to *construct* a story that could, as we have seen, unfold in at least two directions! When the protagonists take all this into account, a "logic of agency" emerges on which they agree. This is how it *had to have* happened. It had to? Isn't it a *construct of the past* according to *our current* understanding of the realities (and language) of the time? Here Owen Barfield's (1988: 67) accusing finger emerges, "Our *present* collective representations arise [and] are determined by ideas and beliefs, rather than [...] resulting from them."

Let's add a digression inspired by contemporary forensic investigative methods—after all, a *story* must eventually emerge here too! I. Dror (2020) draws attention to the often neglected "human factor" that contributes to the shaping of a story—despite all the laboratory results, expert testimony, etc. Yet this is usually not a matter of deception or deliberate distortion: *bias* is the result of the contexts relating to the case itself, the cultural and experiential *milieu* of investigators and experts, experience in evaluating leads, as well as factors such as overworking, knowledge of the circumstances of the case (and outrage at the crime or, conversely, a downplaying of it), the investigator's current health and family difficulties, his or her education and hobbies, the expectations of bosses and the wider public, etc. Perhaps our text should have had a third co-author—from the area of forensic techniques. What we all have in common is the ability to recognize what a lead actually is, the evaluation of such leads, and ultimately the assembling of them into a plausible story. We encourage the reader to read through this filter as well.

Čapek's story can serve as an instructive example in at least two other aspects. First, it shows the long-standing confidence in the productivity of the so-called *direct method* (as opposed to other methods of historical-scientific research) and in the unambiguous objectivity of the source. Secondly, it raises the question of the extent to which it is appropriate to assume that all historical events are *ultimately causally connected.* Archivist Divisek is looking for *clues* that would help him solve the ancient case. However, the fact that he is so firmly convinced of the possibility of reaching a completely objective conclusion, has to do with the older attitude of historiography towards sources and the way of working with them. Nineteenth- and early twentieth-century historiography was convinced that "facts" were directly contained in source texts or artifacts, and that these facts only needed to be "extracted" from the sources by internal and external criticism. Such confidence in the objectivity of the *direct method*, in which *facts are taken* from the sources as if unmediated, was related to what was considered in the nineteenth century to be topics of appropriate historical interest. If we ask, for example, when and in what form the Bohemian principality acquired the hereditary royal title, we can read the date *directly* from the Golden Bull of Sicily of 1212—no interpretation needed!

The fact that the direct method is insufficient is shown, for example, by Zdeněk Kalista, who had great respect for the *givenness of the source.* This can be seen in Kalista's efforts to reconstruct a kind of "ideal source" (Kalista, written 1947, published 2002). At the same time, he had a great sense for the historian's "creative activity", of the role of his consciousness and subjectivity in *understanding* (i.e. not "knowing") history, its individual epochs, events and actors. Kalista points out that we can no longer make do with a merely "direct method" in asking questions that are relevant to the discipline. "This is very different for a historian concerned with spiritual history and its problems. The source may speak directly about what is the subject of interest: it gives a definition of what nobility is, it describes what is to be seen under the words 'grace of God', it defines the limits of the term 'state' and so on, but there are not many sources so unambiguous. People far more think in terms of values and ideas than define them, and they far more often enact and experience an idea globally than formulate it. [...] The historian's narrative does not always grow directly from the source as it does for those who study political history. It requires a distinctly more vigilant spirit if a picture is to rise above the simple, almost naive depiction coming from an old dean of the Prague church, with a piece of his inner life is to be captured, when a light current is to be awakened to illuminate the roots of his fable" (Kalista, 2002: 218–220). Despite being aware of its inadequacy, Kalista did believe in the power of the direct method.

It can be said that as soon as historical sciences began to inquire into facts such as, for example, cultural or social relations, they also had to work differently with the sources. Sourcs not as direct witnesses, but only as *indicators*, as Ginzburg (1989, 2013) puts it. Koselleck's thesis on the "veto of sources", which will be discussed repeatedly in the following chapters, should also be read in this sense. Now, the closeness between the practices of historiography and evolutionary biology becomes all the more apparent. Thus, we have two paradigms of science: *the paradigm of evidence*, which is shared by the historical sciences as well as, for example, by evolutionary biology and the medical sciences; and the 'Galilean paradigm', on which many of the natural sciences are based, but historiographers depend on it as well, as means of identifying their sources.

Carlo Ginzburg (1989), for example, sees it this way, but he is one of the few historians who would find analogies between history and some natural sciences. In history and in the theory of historical sciences, however, the opposite has been frequently claimed, but we will mostly follow Ginzburg in the pages of this book.

Kalista—contrary to what we are trying to show in this book and also contrary to Ginzburg's more recent reflections—believed that there is a stark difference between history and biology. In his works, he compares the practices of the historian to those of the botanist at several points (Kalista, 2002: 76–77 and 96–97). He subordinates the distinction between these disciplines to the more general distinction between history and the "sciences of the present". He clearly does not consider evolution or ontogeny: "The botanist [...] simply *observes* his object, he looks at it, he can ascertain its color and shape with his eyes, he can detect by touch, for example, that a certain plant has rough leaves, or by tasting it he can ascertain the presence of an acid in its leaves, etc." (Kalista, 2002: 96). In contrast, he judges that "the historian's situation is at first sight considerably less easy. The fact to which his interest is fixed [...] does not exist. [...] The historian cannot observe it, cannot stand in the middle of a battle and watch [...]" (Kalista, 2002: 97). The circumstances Kalista argues for, however, appear—as we show—trivial if we view biology and history primarily as a scholarly effort simply to order the multiplicity of traces that are *here and now* along with their *developmental* explanations, and if we ascribe to life a certain semiosis-capable innerness, as exists in the human culture.

Our first concern is to overcome such "trivialities". Divisek, as we have already pointed out, considers that he, as a *scientist*, is obliged to exclude chance. In the heading of this chapter, we point out the idea that existence is an endless series of law-governed events; we now encounter a similar notion among many historians. Čapek, in the character of Divisek, has masterfully captured their vision of the world. Kalista again serves as an example: he writes "Like a huge stream, rolling from infinity to infinity, history appears in the end as a *single flow* in which every event[3] is in its core actually determined from eternity, although for its realization it needs a myriad of intertwining minor and major causes given from time immemorial. We never know when any event actually begins, and we cannot imagine how it will end, for its end shall be the end of history, a hardly imaginable, or even unimaginable *nothing*" (Kalista, 2002: 131).

Thus, Kalista presents the idea that at the very ontic level of the real nature of history we are dealing with an infinity of interconnected, intertwining minor and major causes. It is worth pointing out that, for example, such a notion was rejected as early as 1892 by Georg Simmel (1977). The continuity of history, according to him, is not determined by such an ontic, objectively acting law, but functions only an interpretative continuity, determined by the fact that we organize the diversity of source reports by means of the concepts of development. From the 1960s and 1970s we saw the emergence of narrative critics of history who sought to show that the

[3] Note that for Kalista "event" is undeniably an ontological category. We will see in Chap. 5 that for example in Koselleck's conception of "event" this is not the case.

continuity of history is only a result of a narrative context, that it is an effect of epic formation and not an ontological category (e.g., Veyne, 1984). For many theorists of history, *contingency* has thus become a defining feature of historical reality. This is also an issue that we will repeatedly return to in the following chapters.

2.3 Continuity and Narrative Interpretation

The exuberant beginning of this chapter needs a counterbalance; let us now indulge in a more challenging conclusion, devoted to "world–making through narrative". We deal with *human* cultures, but—as at any places of such context—taking all-life (biosemiotic) stance and contmplating possible links to non-human cultures. Historians Jan Assmann and Frank Ankersmit will accompany us throughout our whole text; here we touch briefly just two of their ideas. In his analysis of ancient cultures, Assmann (2011: 255f) points out that oral and literary cultures differ, among other things, by the phenomenon of *hypolepsis*, which we can perhaps translate in English as "taking up", i.e., it is not necessary to start each time from the beginning. An example of this is the two contrasting statements made in our contemporary culture: "As we said yesterday…" and "As Plato says…" The first utterance presupposes direct participants or witnesses to the discourse; but even they must often first recall and agree on what was "said yesterday". Other members of the culture may have access to what was said only by hearsay, by paraphrase. This is often provided in communities without script (or other recording systems) by specialized persons who, through rigorous training, retain certain parts of the tradition in their memory. With their death (and often before) the memory is lost; while it may be paraphrased by successors, there is no way to verify that the paraphrase is faithful. Thus, oral cultures can only rely on continuity to transmit their memory through recitations, songs, stories, etc.

In literate cultures, writing is a reliable repository (especially if the culture takes care to manage the writings) that can be followed up with further narratives even centuries later; what was written (and preserved) during those centuries can also be followed up. It allows one to recapture unused layers of one's own culture, and for adoption (and translation) from other cultures. Hypolepsis in literate cultures thus allows for a vast network of appropriations, translations, interpretations, and contestations over the references of texts, allowing a culture, among other things, to understand itself differently, and sometimes find new ways, to "rewrite" its own history and its understanding of the world—as well as understanding others, contemporary or ancient. According to Assmann, the aforementioned network manifests itself in three ways, and he provides casuistics of archaic civilizations from first millenium BCE. Egypt, the cultures of Mesopotamia, and the Jews (each on their own ways, as will be commented below; see Box) established a canon at the foundation of their culture, a list of texts that cannot be altered, subtracted from, or added to. "Canonical texts cannot be changed—this marks the crucial difference between them and the stream of tradition. They are sacrosanct, and must be handed down word for word"

(Assmann, 2011: 78). Of course, there is a catch. For such a culture to function in the long term, it is necessary to establish a specialized institution of scholars who provide the rest of the population with an interpretation of specific passages of the canon and their meanings, as a picture of the world that the population can then take as its own at any given moment. The Greeks, by contrast, had no such strict canon, so the texts could be studied and interpreted by all literate members of the population—even at the cost of disputes, contradictions, schools of interpretation, etc. Christian cultures then represent a symbiosis of Jewish and Greco-Roman approaches to the written word.

Box

Three paradigmatic civilizations as characterized by Assman can help, we believe, to understand some life strategies:

1. Egyptian civilization is concentrated on the canonic structure of the world: it is mirrored by styles of architecture of temples, hieroglyphic texts in them, pictures and plastics—all is frozen, copying endlessly the presumed properties of the world; nothing should change. No written description (manuals, recipies, commandments, etc.), all is displayed iconically.

   Some cellular structures are also depending on te design of existing structures: They cannot get assembled *de novo*, in spite that the building materials are available. Such examples are, e.g., biological membranes, kinetosome and flagellae, and, of course, the cell as a whole.
2. Civilization of Israel, in contrast, is build on canonical texts *prescribed* or *negotiaded* with the deity: no change is allowed in the canon, but as time goes, its interpretations abound, in order to abide the assigments.

   Parallel to this understanging is life dependend on strict observations of *codes*: living beigs are alike programmed machines, with written commands on one side, and bodily structures capable of understanding such commnds and performing accordingly. Such strategy is common in short-living organisms, or in tasks that need to be carried out quickly and seamlessly (as, e.g., development of early embryos).
3. Greeks had no such canons, their civilization was based in text freely available to interpretations, and producing their variants, or even brand-new texts. All such texts entered the hypoleptic network of mutual influencing, from which different schools of describing the world arose.

   As highlited at many places of our text, this model is the true analogy taken up by the biosemitic endeavor, i.e. living beings, species and even biosphere work like self–interpreting entities, in this process defining their present place in the world.

We should add, of course, that no such models should be taken absolutely, neither by historians or biosemioticians: all work in a mutal interplay, with this or that traits accentuated at particular places and times.

Without knowing anything about Assmann's hypolepsis, Markoš & Švorcová in their book *Epigenetic Processes and the Evolution of Life* (2019) propose a similar scheme for evolution. Genetic notation is very much a written norm, a kind of "canon" that should not be changed too much. This canon can nevertheless be interpreted in different ways along different life lineages. Again, as with human cultures, it allows for returns to forgotten sections of the texts ("anamnesis") and/or their reinterpretation. This is best seen in ontogenesis, where different stages of development can "read" their written genetic legacy in different ways. Even more interesting are the new readings of preserved parts of the written heritage. Another tool for new interpretative settings is horizontal gene transfer, i.e. adoptions of scripts from other lineages (other versions and additions to the canon that the adopter can "read" despite the differences). In all such cases, they must somehow be understood, i.e., interpreted. These are often ancient norms, and thus a link to all inhabitants of the biosphere (this is joined by the other universal forms of behavioral co-communication characteristic for Closure 1 and for certain forms of Closure 2). An analogy of the biosphere so conceived is, e.g. Christian civilization with its multitude of denominations and sects, nations and cultures.

Back to human cultures again: "As some texts disappear, some enter the flood, and other are extended, shortened, revised, or anthologized in ever changing combinations. Gradually, different structures are formed from the center to the periphery. Some assume central importance because they are copied and quoted more often than others; thus, they eventually become classics that embody normative and formative values. The school of writing played a vital role here; it created an institutional framework for the copying, circulation, and archiving of texts, thereby ensuring that the texts and their values remained present and accessible. In this way a 'great tradition' emerged that created a repository of knowledge and, more importantly, education that extended back over hundreds and then thousands of years" (Assmann, 2011: 76–77).

Turning now to Ankersmit (1994), for whom forms of care for past heritage emerge from a somewhat different yet actually very similar angle; again, let us not forget that the author speaks of human cultures and is concerned with narrative interpretation of the past. He formulates six theses (and several sub-theses and sub-sub-theses) of the narrative interpretation of history that can provide the reader with a guide for the later chapters of our book; if we seek to extend these to other life-forms in the biosphere thus bring them into the focus of biosemiotics, it is to our credit. The basic theses (somewhat paraphrased) are as follows:

1. Historical narratives are interpretations of the past.
2. Narrativism accepts the past as it is, in the form of a tautology: it accepts what is unquestionable about the past. What is unquestionable is a historical fact. Both senses of the last statement are true.
3. Narrativism is the modern heir of historicism: it recognizes that the fundamental task of the historian is interpretation.
4. Interpretative language is not objective language.

5. Statements of historical narrative always have a dual function: 1. to describe the past; and 2. to define or single out a specific narrative interpretation of the past.
6. The roots of historicity are deeper than anything proposed by contemporary historiography or philosophy (of history).

We will not deal with the sixth thesis, but stop shortly with the first five; our selection of sub-theses below represents about one-fifth of the whole corpus.

Theorem 1.2.1 states that "Scientific theories are underdetermined since an infinite number of theories may account for the known data; interpretations are underdetermined since only an infinite number of interpretations could account for all the known data." Here we just say at the outset that Divisek and Mejzlik will never be sure of their solution (if they would be willing to admit it is open to question), even if many more clues for their story accumulate over time, and even if all support it. Related to this is sentence 1.3: "Interpretation is not translation. The past is not a text that has to be *translated* into narrative historiography, it has to be *interpreted*". Sentence 1.4 and its development is then a reminder that even historical time need not be taken at face value—it is a recent invention of Western civilization. Finally, sentence 1.5 notes that as late as the 1970s, scientism still ruled the philosophy of history; to abandon it now is not to jump to the opposite extreme and declare historiography to be fiction.

The second thesis says as much, 2.1: "It is necessary to distinguish between historical research (a question of facts) and historical writing (a question of interpretation). The distinction is similar, though by no means identical, to the distinction in philosophy of science between observation statement and theory."

The third thesis, which refers to historicism, may surprise the reader, but gets close to our biosemiotical project. However, Ankersmit explains (emphasis ours): "3.1 Interpretations strive for the unity that is characteristic of things. 3.1.2 Historists attempted to discover the essence, or, as they called it, the *historische Idee,* which they assumed was present in the historical phenomena themselves. Narrativism, on the contrary, recognized that a historical interpretation *projects* a structure onto the past and does not *discover* it as if this structure existed in the past itself. 3.1.4 Insofar as the notion of plot or intrigue is suggestive of a structure or story present in the past itself, this notion is an unwarranted concession to historist, or narrativist, realism." It follows then (3.2) that interpretation is not a reflection of the past and is not bound by the rules of translation (see also 1.3 above), hence: "3.3 Narrative interpretations apply to the past, but do not correspond or refer to it (as [parts of] statements do)." This is just an elaboration of the idea that information both archival and factual is different from its position in the story itself. The language of narrative is autonomous with respect to the past; there is no fixed relationship between interpretation and the past; such a claim would be a categorical error.

The fourth thesis is perhaps the most developed, but it merely refines what has been discussed in the previous theses; here we highlight only some of the conclusions. First, it addresses the relationship between language and the world: "4.1. Narrative language *shows* the past in terms of what does not *refer* or *correspond,* to parts or aspects of the past. Narrative interpretations in this regard resemble the models used by fashion designers for showing the qualities of their gowns and dresses. Language is used for showing what belongs to a world different from it." And we approach the final message: "4.7 Narrative interpretations may have proper names (like the General Crisis of the Seventeenth Century, the Cold War, Mannerism, or the Industrial Revolution). Mostly, however, this is not the case," but also "4.7.4 Narrative interpretations have no existential implications (for example: *the Industrial Revolution* is not a vast impersonal force in and of historical reality, unnoticed and undiscovered until 1884 when Arnold Toynbee wrote *The Industrial Revolution in England,* but an interpretative instrument for understanding the past)."[4] And here comes a very important observation for us: "4.7.5 Nevertheless, if a narrative interpretation goes unquestioned for a long time, is accepted by everybody, and becomes part of ordinary language (thereby losing its historiographical nature), it may turn into the notion of a (type of) thing. A *narrative thing* (see 4.4) has become a *thing in reality.* This is how our concepts of (types of) things originate. Typification procedures decide what is still merely interpretative and what is real; there is nothing fixed and absolute about the demarcation between what is interpretation and what belongs to the inventory of reality." Peirce's notion of habit expresses the same thing in semiotics.

References

Ankersmit, F. R. (1994). Six theses on narrativist philosophy of history. In *History and tropology. The rise and fall of metaphor* (pp. 33–43). University of California Press.

Assmann, J. (2011). *[1997] Cultural Memory and Early Civilization: Writing, Remembrance, and Political Imagination*. Cambridge University Press.

Barfield, O. (1988). *[1957] Saving the appearances. A study in idolatry*. Wesleyan University Press.

Čapek, K. (1994). *[1929] Tales from two pockets*. Catbird Press.

Depew, D. J., & Weber, B. H. (1996). *Darwinism evolving. Systems dynamics and the genealogy of natural selection*. MIT Press.

Dror, I. E. (2020). Cognitive and Human Factors in Expert Decision Making: Six Fallacies and the Eight Sources of Bias. *Analytical Chemistry, 92*, 7998–8004.

Eco, U. (1989). *[1967] The open work*. Harvard University Press.

Ginzburg, C. (1989). [1980] Clues: Roots of an evidential paradigm. In *Clues, myths, and the historical method* (pp. 87–113). Johns Hopkins University Press.

Ginzburg, C. (2013). *The Night Battles: Witchcraft and Agrarian Cults in the Sixteenth and Seventeenth Centuries*. Johns Hopkins University Press.

Kalista, Z. (2002). *[1947 and the estate] Cesty historikova myšlení [Strolls of historian's thinking]*. Garamond.

Markoš, A., & Švorcová, J. (2019). *Epigenetic processes and the evolution of life*. CRC Press.

Simmel, G. (1977). *The problems of the philosophy of history. An epistemological essay*. The Free Press.

Veyne, P. (1984). *[1971] Writing history: Essay on epistemology*. Wesleyan University Press.

[4] The same may hold for natural selection in evolutionary biology.

Chapter 3
About Things and Objects

Abstract Things are elements of the world or of the innerness of living beings; they may or may not be independent from ou minds. Interpretation reduces some characteristics of things into objects; things will thus be reduced to objects: "unknown object" is an oxymoron. Interpretative circles further specify the meaning of objects, sometimes ending in habits. Latin civilization as an originator of the semiotic theory–a tool of world interpretation. Basics of the semiotic theory in the Modernity, with signs and relations in its centre: Augustin, J. Poinsot, C. S. Peirce, J. Deely, and U. Eco. Semiotic as a tool for humanities: broadening its scope to all life. Organic codes as an attempt to develop scientific interpretative theory. Language and the arbitrariness of signs. Relationships of semiotics and historical sciences: difficulties with reified objects as nations, cultural forms, moods, etc.

Keywords Thing · Object · Habit · Code interpretation · Signs and relations

> We have seen that the essential and characteristic achievement of all symbolic form—whether of language, myth, or pure cognition—does not lie simply in receiving given material impressions (which *in* themselves possess a fixed and definite character, a given quality and structure) and then grafting onto them, as though from outside, another form originating in the independent energy of consciousness. The characteristic achievement of the spirit begins much earlier than this. On sharper analysis even the apparently "given" proves to have passed through certain acts of linguistic, mythical, or logical-theoretical apperception. Only what is *made* in these acts "is"; even in its seemingly simple and immediate nature, what is thus made proves to be conditioned and determined by some primary meaning-giving function.
>
> Ernst Cassirer, 1955: 94.

Let's leave Mr. Divisek and Dr. Mejzlik at their praiseworthy activities. But we must—with thanks to Karel Čapek for masterfully capturing their style of thinking—question their certainty a little. We will do so with the help of the theory of semiosis, to which we will refer many times throughout the book. Here we will

A. Markoš, J. Horský, *The Seeds of Forms and the Forms of Seeds*, Biosemiotics 12, https://doi.org/10.1007/978-3-032-16131-4_3

explain the theory of the sign (we are inclined to its triadic conception), but only selectively, insofar as it is interesting from the point of view of biology, biosemiotics and historical sciences.

3.1 Things and Objects

Let us return to semiotic, or biosemiotic, depictions. Deely (2010 and in other books) gives us an amazing account of the history of Latin civilization in terms of the development of semiotics. Semiotics, he argues, is one of the fundamental inventions and contributions of this civilization, and Deely unravels its struggles from the time of Augustine (354–430) through the scholastics to the first half of the seventeenth century when the Latin Age passes the baton to modern thought. Latin civilization was primarily concerned with the history of Christian salvation, with the characterization of God, the world and man in it; and to a layman today, unschooled in theology, all those problems and disputes provide no clue to anything. Deely offers a narrative that brings a different perspective to the Latin centuries, a new interpretation of the world. As he sees it, in the second half of the nineteenth century the founder of semiotics, Charles S. Peirce studied in detail then dismissed Kant and his followers before picking up Latin thought and conveying a forgotten doctrine in a form adapted to the modern age (and thus lays the foundation for the coming post-modern thought).[1] We now turn to a brief survey of Deely's interpretation of events, as it provides a wonderful opportunity to understand all earthly life. (Again: such a broad scope was apparently *not* his own intention—he considered semiotics a knowledge of *human* understanding.)

Augustine of Hippo, the principal founding father of the Latin tradition, introduced into Latin thought a novel concept of sign (*signum*). He was probably unaware of his breakthrough achievement: he may have thought he was merely translating the Greek concept of *semeion* (his knowledge of Greek was apparently rather poor). However, while *semeion* refers only to natural manifestations (for example, fever is a manifestation of illness), *signum* also refers to language and to the human innerness, and these contribute decisively to understanding. For Augustine, human innerness is guided by the Word of Christ (*verbum cordis* or *verbum interius*). Thanks to His guidance, humans can distinguish between perceptions coming from the outside, and therefore between good and evil: we can *know* the way to salvation. Better still: by our own active effort we can *reach* this path. Human innerness thus plays a key role in knowing the world and living righteously in it, and only through innerness can we not only use and know the world, but also enjoy it. In Augustin's crucial work *De doctrina christiana* we read:

[1] Not to be confounded with postmodernity from the end of the twentieth century: in Deely's view this moment represented a mere climax of Modernity, not a Post-Modern age.

> All teaching is teaching of either things or signs, but things are learnt through signs. [...] There are other sings whose whole function consists in signifying. Words, for example: nobody uses words except to signify something. From this it may be understood what I mean by signed: those things which are employed to signify something. So every sign is also a thing, since what is not a thing does not exist. But it is not true that every thing is also a sign (Augustine, pt. 1.4, 1997: 8).

Our ideas, experiences, memory traces, etc. reflect things—those existing in the world, but also those in our mind—and they can also act as *signs*, i.e., denote something else. This basic concept has been refined and added to over the centuries, and the doctrine culminated in the late scholasticism of John of St. Thomas *a.k.a.* John Poinsot (1589–1644). Deely dedicated to Poinsot (among others) the first two parts of an unfinished trilogy (Deely, 2008, 2009b), as well as the bilingual (Latin-English) *Tractatus de signis* (Poinsot, 2013), which consists of Deely's extracts from Poinsot's much more extensive work, along with Deely's own commentaries.

The semiotic teaching may have peaked in Poinsot, but it no longer influenced the world, which went in a different direction, inspired by Poinsot's contemporaries such as Galileo and Descartes. This remained until the second half of the nineteenth century when C. S. Peirce began to study medieval scholasticism and build a new doctrine, semiotics. In Deely's reading, semiotics has run in a parallel to modern philosophy as it unfolded in Kant.[2]

Semiotics is, of course, primarily a human-centered science, and it is not in our power to cover all the facets of the immense legacy of Peirce and his followers; we will therefore stick to those aspects of the doctrine that fit into Deely's story. This will, we believe, allow us to apply semiotics to all living beings without getting entangled in the subtlety of Peircean nuances. We must at once point out that Deely opposed such an extension to all life: rather reluctantly he added the so-called "higher" animals to the teaching aimed originally at human affairs, but he never would accept biosemiotics, as we do below; *caveat lector*! The following is a summary of Deely's semiotics, which follows in that tradition; except for differences in terminology, it is, we believe, similar also to the teachings of Umberto Eco.

Deely is working with the basic notion of *relation*, one of Aristotelian accessorial categories.[3] Relation is always a category involving two *things*. Things are independent of our consciousness, so they represent *subjects*. The establishment of a relation of such a pair of things requires a cognitive (i.e., living) being, for only such a being has memory and experience (and the attention that goes with it), and thus is able to interpret perceptions aroused by things (both animate or inanimate; external or internal). By means of a relation which is suprasubjective (i.e., it transcends both subjects involved), the cognitive being interprets its perception of a thing by constituting an *object*—in accordance with its own experience. The thing in its totality cannot be known, but it can give itself via various signs (carried by the thing proper,

[2] Let us point out in passing that Deely regarded modern philosophy (founded by Kant) as solipsistic, which is to say a blind branch of knowledge: he assumes it will be overcome by semiotics.

[3] According to Deely, all Aristotelian accessories can be reduced to relation.

or by other sign-carriers referring to it), and so it is known via various objects. In the preface to his book *Purely Objective Reality*, Deely states:

> Anything, to be an object, must exist in awareness. *To be known and to be an object: the two are the same.* But what is it, then, that an object is as such? For not everything needs to be known (at least not by any finite intelligence) in order to be at all. To be is sometimes, yet not always, more than to be an object; but to be an object is always a form of being. *And every form of being which does not start out as an object yet can become one in the right circumstances.* As we will see, the question is one of principle, not merely of fact. For not only are some things unknown to us, we have every reason to expect that we will never know everything that could be known, though what it is that we will never come to know varies from individual to individual and group to group. (Deely, 2009a: 8; emphases ours).

In many places, Deely emphasizes explicitly: the "unknown object" is an oxymoron, because only objects are communicable and learnable in any given community and at a given time; hence they can become a common property of the community.

The formed object (i.e., the model of the thing) is projected back onto the thing and thanks to this we may be able to recognize additional signs on the thing, thus further refining our ideas about it. Such a circularity between the thing, the carrier of the sign and the interpreter can go on forever, or after a while the process stops, and the object is declared—for the moment—to be the thing itself. A similar diction can also be found (with minor differences in terminology) in Umberto Eco's writings where he interprets Peirce's doctrine of how we know something.[4] An older insight into the dynamics of symbolic acts is provided by Ernst Cassirer (1946). In his thinking, individual perceptions and feelings become, by acts of symbolic formation, stabilized, repeatable, and communicable. The "object" is, on the one hand, a reduction of the multiplicity and diversity of a certain group of feelings or sensations, but on the other hand, it is also a certain "extension" (supplementation or enrichment) of them, since in symbolic formation it introduces a certain more general content that *was not (or did not have to be) in the original group of represented sensations and percepts*.[5]

In other words, we (living beings) interpret signs and *construct* the object: the world as we perceive it is subject to endless semiotic interpretation (construction) through objects (Fig. 3.1). If this process ceases, the world empties into an ideology that *knows* how the world works. But we should repeat again: the entire teaching of

[4] "A Dynamical Object ['thing' in Deely] drives us to produce a *representamen* [the carrier of the sign], in a quasi-mind this produces an Immediate Object, which in turn is translatable into a potentially infinite series of interpretants and sometimes, through the habit formed in the course of the interpretative process, we come back to the Dynamical Object, and we make something of it. [...] As soon as we get back to the Dynamical Object and start to speak about it again, we are once more at the point of departure, and so we have to rename it using *another representamen*. [...] The Dynamical Object is Something-that-sets-to-kicking-us and says 'Talk!' to us—or 'Talk about me!' or again 'Take me into consideration!'" (Eco, 1999: 13–14) Things invite us—living beings—to explore the world of things through signs and objects.

[5] Similarly, the object is an "ideal type" (Weber, 1904) or "plot" in the framework of a "historical narrative interpretation" (Ankersmit, 1994; Ricoeur, 1984) that brings into parts of (past) socio–cultural reality, into incidence, a certain more general relational or event pattern that was not (or need not have been) previously there.

Deely, Eco, Cassirer and others is elaborated for the needs of the humans, "we living beings", it is *our* extension. In Deely's eyes, for example, some animals do recognize objects, but they have no idea that behind the objects there are things—so their world (umwelt) is a kind of (unchanging?) "ideology", i.e., "objective reality". Other beings are apparently not worthy of Deely's attention at all—they abound in sensitivity, but not in receptivity—i.e., they do not interpret the sensations, they merely function as programmed machines (with Barbieri's proviso that they are not machines after all—see 3.3 below). The whole effort of our text is to arrive at the presupposition that a living being is and must be an interpretative system, for us to be able to talk about biosemiotics and Closures of type 1 and 2, without forcibly cramming all this into experimental science, where it has no place. A useful factual background for these considerations is provided by epigenetics (Chaps. 10 and 11).

3.2 Signs and Relations

Let us try to illustrate our Deely–inspired reasoning, represented by the diagram in Fig. 3.1: it is in particular a maximally simplified synthesis of the teachings of Peirce, Deely and Eco, using Deely's terminology (here as well as throughout the rest of this book).

1. The world is full of *things*, and they exist "for themselves", regardless of whether anyone recognizes, deals with, or perceives them against the background of the world. Let the stars and moles be our examples and assume that we know nothing about their existence.[6]
2. We somehow learn about things through perception (real or imagined) of their manifestations. This is always a reductive effort because we can never perceive a thing in its fullness. The thing is somehow given to us;[7] what we perceive and what we record is a kind of reduction of it on our part to *an object*, **O**. It is usually given to us through signs bound to a *sign carrier* (*representamen*, **R**). Such sign carriers in our two examples are the point lights in the sky and the molehill, and these will induce in the interpreter the assumption of the existence of objects, which he will call, for example, *a celestial light*, or *an underground creature.* "We" approach the world as interpreters, seeking to interpret it as it appears to us in any given moment.

When we analyze the sky, focusing on the point sources of light, (1) we naturally exclude the Sun and Moon from consideration. Then we will classify the other

[6] Eco's impressive study (Eco, 1999: 127f) recounts fictional difficulties that the Aztec messengers to their emperor might have had when they announced the approaching Spanish cavalry: how to explain to Montezuma what a horse is?

[7] The manner or degree to which we are given things is a matter of contention and debate between proponents of different paradigms. They are primarily concerned with how the construction of an "object" is affected by the form, even the existence, of "things".

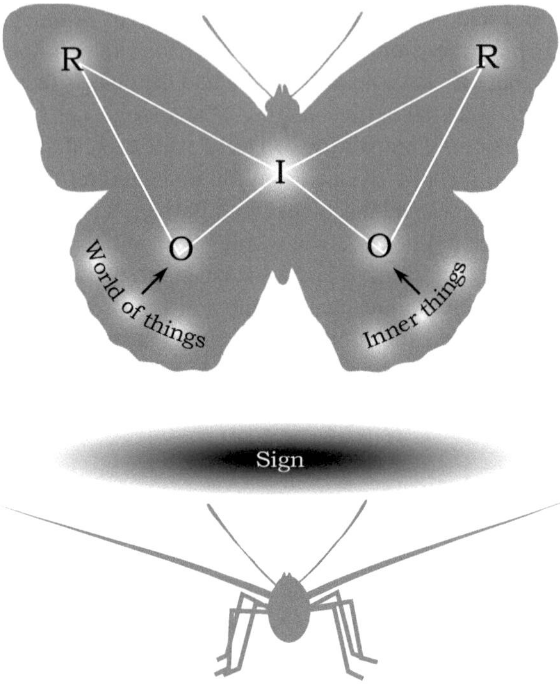

Fig. 3.1 Diagram of the semiotic process according to Deely (in our adaptation). On the left the external world, on the right the innerness of a living creature. Things (external and internal) are given in the form of objects O, *via* carrier of signs R, such that the interpreter I is able to perceive it. The interpreter integrates the inputs and puts them into interrelationships, which enables it to render the object better into more details. The two triadic relations are irreducible to their components, and their superposition produces a *sign* that is thus "suprasubjective" (i.e., not bound to the subject's material properties)—this is what the projection of the diagram at the bottom of the figure attempts to illustrate

lights (possibly naming the more prominent ones): (2) those light sources moving simultaneously with the whole sky shall be called stars; (3) the small number of sources moving independently of the others—these are the planets and occasional comets; (4) and finally there are fast-moving point sources of light. By further circling in the Object–Representamen–Interpretant cycle we refine the characteristics of the objects under study. Thus, we break up group (1) and the Sun surprisingly finds itself in group (2), i.e., as a star, even though it does not fit the original definition of a *star* object. After all, it is not a point source, but its motion in the sky is indeed somehow related to the motion of the sky, even if in a different way than is the case of other stars. At the same time, it is not a planet. Further analysis will show the surprising fact that the Earth, while not being a light source (let alone a point source), is a planet, while comets are not planets. The Moon surprisingly stays as the sole and unique object of group (1) as a *satellite of the Earth*; only later will the successively discovered satellites of other planets enter this group. We leave the analysis of group (4)—meteors, landing planes, artificial satellites, comets, planetesimals, etc.—to the reader. An object initially recognized as a "celestial light" becomes (with the exceptions of aircraft and man-made satellites) a "celestial body" and receives any number of additional attributes—depending on its membership in one group or another.[8]

Now we come to a claim that Deely *does not make*, indeed he would probably protest. We take it as given that the interpreter needs not be limited to a human being, yet *must* be a living creature, as we have characterized it with the help of the combination of the two Closures. Still, let us first consider a man who was born into a culture that accumulates knowledge—and constructs objects based on celestial phenomena over thousands of years-long of observations the sky. The innerness of the human interpreter (recall, he/she is one of the spatio-temporal occurrences of Closure 2, made up of, among other things, human cells) forms the virtual counterpart of what is given to him through his senses from things of the external world. In school, or even before, he is given a much deeper knowledge of cosmic objects, derived from observation, calculation, introduction of concepts such as celestial body, interstellar space, gravity, trajectory, orbit, relativity, constellations, etc., so that apparently no one today undertakes the mind exercise indicated above. The child gets most of the space objects categorized right in the textbook, but also learns a lot of other properties of *the things behind* the objects so constructed: this includes the mass of various bodies, their surface temperature, the origin and evolution of the whole universe and the supposed things in it, etc.

These things constructed in the mind and their corresponding objects—*verbum interius*—are part of the interpreter's innerness, and he/she constructs the image of the world from both worlds: the external and the internal, moreover, he/she can also change it. And now a very important fact: things and objects are also the product of "collective" labor within the limits of various types of Closures 2, and the results of such labor can also change (see Chap. 8).

Even other living beings—non-human inhabitants of the biosphere—can become interpreters of cosmic phenomena, but from the point of humans such grace usually ends just with a minor concession—it is extended only to animals endowed with brains (vertebrates, some worms and arthropods, perhaps cephalopods). This is also Deely's point: he judges that, unlike humans, these animals only recognize objects (and produce inner objects): they don't care that there are things behind the objects that *somehow* give themselves through the objects.[9] Other forms of life have hardly

[8] Deely gives a similar astronomical example: "Before we could inquire into the rings of Saturn, we had to become aware that Saturn had rings. And before we could realize that Saturn was not the only planet in the solar system with rings (as was believed and taught in the schools throughout the 19th and for most of the twentieth century), we had to discover that some other planets had rings as well. And so on. Science does not advance merely by objectifying the physical environment, but rather by objectifying the subjective structures as such to which we initially gain access by the simple means of our bodily senses, but which we come to know well beyond what those senses can present through those extensions of sensation and perception that we call scientific experiments, theory, and research" (Deely, 2009a: 9–10).

[9] We will not examine whether this is merely the anthropo-chauvinism of scholars like Heidegger (1995); his investigation led to the description of animals as "world-poor" critters, because they

received any attention from semioticians, and it is only in the biosemiotics endeavor of the last quarter of a century that we discover that signs are common to all living things—which also this book attempts to do. We assert that the interpretative process is always a combination of the recognition of external stimuli *and* the contribution from the inner world of memory and experience (evolutionary, personal or collective), although no mental processes need be assumed. Examples of non-human astronomical interpreters include migratory birds orienting themselves according to star constellations; plants perceiving the length of the day and adjusting their flowering in accordance with other factors; or the polychaete *Palolo viridis* controlling its reproduction according to the phases of the Moon.

Now our second example—the mole. Unlike the celestial bodies, as a living body it *has* an innerness and therefore fits the diagram in Fig. 3.1. Most of us never met one, we know of them thanks to pictures, or perhaps of cadavers occasionally found on the field roads. Its representamina—molehills—are known to everyone, and in Central Europe, there is no mistaking them: it's the mole who made them. Again, we know much more about the *Talpa europaea* object from school. We have been taught about its way of life, its eyesight, its mammalian nature, its food, many people can place it in the zoological system, etc. All this is again based on our "innerness", cultural memory and experience that has reduced *the thing* "mole" to an object item in a textbook such that we can compare it to other creatures inhabiting the underground; all this helps us create in our minds an object—the mole. At the same time, we do not have access to how the mole itself perceives the world, to *its* interpretation of the world.

All this shows the *relationship* between the elements of the interpretative process, both external and internal, and what remains is to describe the ball or cloud that hovers over the O-R-I interpretation triangles in Fig. 3.1. Deely calls it *relationship*, and its shape responds flexibly to changes in external and internal elements, integrating them into a momentary "picture of the world"; the interpreter, then, attributes *meaning* to the recognized picture accordingly. Deely therefore calls the relation a *suprasubjective* entity: it is neither an object of the interior (subject) nor of the external world (you can't point a finger at it).

To quote Deeely again: "What we call signs are signs only when and insofar as they occupy the determinate foreground position of representing [...] within or 'under' a triadic relation suprasubjective as relation: so that the *signs,* strictly speaking, consist in the triadic *relation itself* rather than the foreground element 'in' or 'of' that relations: *this* realization, establishing the Way of Signs as presupposed for any knowledge whatever of 'things', came to be established among the Latins only as the Latin Age reached its end, in the very lifetime of Galileo and Descartes" (Deely, 2010: xxxvi).

Compare our two examples. The thing "celestial body" has an independent existence, but its relations with the world are entirely determined by the timeless physical "laws of nature": its behavior is imposed on it, it is not a matter of interpretation. The thing "mole", on the other hand, carries with it the memory and experience of

cannot bring their interpretative efforts "to the things themselves."

4 billion years of life leading to a particular individual, and this makes *that* individual the interpreter of the world, and *its* interpretation is unique—even compared to other individuals of the species *Talpa*. All the more so because its world, so constructed, is different from the human one and we have no access to it; we can only roughly estimate the world of the mole through the objects given to us.

We now interrupt our exposition of Deely's doctrine with an excursion into the theory of organic codes by Marcelle Barbieri, who is a passionate opponent of Deely.

3.3 Organic Codes and Disputes About Interpretation

Marcello Barbieri (e.g., 2003, 2024) is one of the most prominent and tireless promoters of "scientific biosemiotics" and later (also scientific) "code biology". His credo is "Everything about life is knowable, testable and provable by biological (read scientific) methods". And one more quote, often repeated in different versions, but very difficult to decipher: "Organic codes are not metaphorical but real" (e.g., 2009b). *Testability* and *knowability* are keys in this regard, and thus proof of *the physical* nature of the life's manifestations: we must not allow these processes to have properties that can be explained only "mystically", or even just "metaphorically". Barbieri is content with objects, he does not acknowledge "unknowable" things, as do Deely and Eco. Most of all, he is irritated by the transcendent, suprasubjective nature of relations that Deely promotes. And of course, he does not recognize the interpretative capacities of the living things except human beings themselves. Hence his attempts to dismiss the possibility that a living being (other then human) might have the capacity to *decide*. Let us follow Barbieri's argument through this lens as well; just recall that this is *molecular* biology, which does not need any Closures. Yet, life differs from chemistry, but only in that through evolution it has developed a set of various *codes*, and these, being historically conditioned, cannot be derived from chemistry.

Barbieri developed his ideas in his book *The organic codes* (2003) and in many books and papers that followed (the latest we know is Barbieri, 2024); however, we will try to illustrate the stream of his thought that concerns us with a shorter article, *A Short history of biosemiotics* (Barbieri, 2009). The article mentions the astonishment caused first by the discovery of the storage of genetic information in DNA and then by the discovery of the genetic code itself, and puts it in the context of the slightly older idea of the so-called von Neumann probe: "Von Neumann showed that a self-replicating system capable of open-ended evolution must necessarily contain a description of itself, and such a description must be categorically different from the construction it controls. […] According to von Neumann, in short, an evolvable self-replicating system must be a *physical system controlled by symbols*" (Barbieri, 2009: 223; emphasis the author). The "construction" or "system" is thus governed by instructions ("symbols"), and these *will*, among other things, *cause* it to copy these instructions. As they contain also the full description of the system, they are also passed onto the "offspring". This brings us to biological systems and their evolution: when instructions (DNA) are copied, mutations can creep in and the implementation of the program can sometimes be flawed (an imperfect "understanding") and a modified "system" is created. Both kinds of mutations result in new "phenotypes".

Commentary: the instructions mentioned above are for a *cell*, and a cell is a body: either a descendant of a mother cell, or a result of the fusion of two (sometimes more) different cells (each with a unique load of instructions)—in neither case is it a product, despite Barbieri's definition of life as "artifact-making" (e.g., Barbieri, 2015: 11). Let us add that the body is "set by tradition", by a chain of ancestry. And the instructions themselves have not been inserted ("uploaded") into the body from the outside; they are always an integral part of it, and a very complicated part. Moreover, unlike instructions, the body of the cell is also influenced by the environment, and its further development may respond to environmental changes as well as the cell's perception of them. So, the question is whether or not the cell (or the developing embryo) does or does not extract *meaning* from the message, i.e. whether it will behave like a self-copying programming machine *sensu* von Neumann, or whether it *reads* and relates to its genetic heritage also on the basis of other sources of memory and experience.

Recall once again that Barbieri is a strict proponent of the scientific method, so everything that makes life different from the usual physical models *must* be testable and scientifically explainable. Just as the existence of a steam engine does not follow in any way from the laws of physics, but once the engine is here, its operation is explicable in minute detail with their help. In this way, we can also easily explain the physical principles of programmed machines like jacquards or computers, but we cannot (yet?) propose a model and physical explanation of *reading*, i.e., not the implementation of a program, but the interpretation of a text, or characters in general. The question comes back again and again: are only humans (or perhaps only animals equipped with brains) capable of interpreting the world, or is the ability to interpret coextensive with life in general? If the second alternative is true, we must consider that (bio)semiotics as a science of interpretation does not belong to experimental sciences, it belongs to the humanities, and its methods and models should be constructed differently. Barbieri, as a scientist, did not accept the fact, and therefore strives for a "scientific biosemiotics", fully absorbable by science and applicable to the interpretation of living beings. This semiotics must then be *devoid of interpretation*, since the opposite would be, he says, an admission of vitalism. Hence his years of disputes with semioticians—but also with scientists.

But let us finally come to Barbieri's selected article from 2009. The basic thesis is stated right in the introduction. "Biosemiotics is the synthesis of biology and semiotics, and its main purpose is to show that semiosis is a fundamental component of life, i.e., that signs and meaning exist in all living systems" (Barbieri, 2009: 221). This is proclaimed by all biosemioticians indiscriminately, but Barbieri immediately trumps it with the statement: "Nature is governed by objective immutable laws, whereas culture is produced by the mutable conventions of the human mind" (Barbieri, 2009: 222). Hence, it is understood that interpretation is a feature of human culture alone. The contrary view that any community of living beings (i.e., not a random accretion, but the situation of Closure 2) could constitute a culture is not considered (for a polemic with this view, see Markoš, 2002). If biosemiotics is to be accepted into biology, Barbieri argues, then it must be defined differently, and he comes up with a three-stage model of semiosis as it has gradually emerged in evolution (see below). The main emphasis is on the existence of *codes*, translators of information from one "world" to another. Codes cannot be derived from physics, they are a unique *historical* convenience of the living world, they have been *negotiated*, they are *organic conventions* ("organic" as opposed to human, cultural ones). But do "conventions" belong to objects, can objects negotiate or bargain?

Barbieri then advocates three levels of semiosis, as they have gradually emerged in evolution: "Life is essentially about three things: (1) it is about manufacturing objects, (2) it is about assembling objects into functioning structures, and (3) it is about interpreting the world" (Barbieri, 2009: 234). The impression is given that the role of semiosis (semiotic acts) increases or gains in quality or importance over time. But this, we argue, degrades the function of semiotics—after all, interpretation is a *prerequisite* of semiotic processes, not their culmination in humans, or some "higher" animals. Let us have a closer look at the three steps.

The first stage—*semiosis of production* basically means "production" of nominabilia.[10] So proteins exist according to instructions stored in DNA and translated through the genetic code, and then the proteins take care of the rest of the "production process". The first organisms were programmed almost to a fault by genes; today, cells and "lower" multicellular organisms are at this level. They are, according to Barbieri, unable to create their own internal representation of the world.

At the second stage, therefore, *signaling semiosis* comes into play—this is the enormous amount of signaling, regulatory pathways that we find in all animals (it is not clear from the depiction whether this includes, for example, plants—whether they do not belong to the previous, first stage). These pathways ensure their survival by reconciling vital functions and stimuli from the environment and from within the organism. For twenty examples of this type of code, see the special issue of the journal *BioSystems* (164, Issue 1, 2018).

These stages are both "not metaphorical but real", which pretty much means just that their components can be "pointed at", *scientifically* described, predicted and tested: i.e., these are essentially physiological processes under a different code name (see also Chap. 10). Interestingly, Barbieri nevertheless does not equate living beings and programmed machines. This is because life, he argues, is a result of evolution, and its codes evolved from within, by "negotiation", whereas computers and robots got their code systems from without, brought about by humans. Here, then, is the neuralgic point of the theory: a negotiation that needs no agent, a kind of as if automatic self-negotiation, if we understand his argument well.

Only the third stage is *interpretative semiosis*, i.e., what other semioticians consider semiosis *sensu stricto*. It is mainly associated with the faculties of thought and

[10] Biopolymers (proteins, nucleic acids) are unique, not further reducible, i.e., they "have a name".

language in humans, with perhaps some license for "higher" animals. Here, of course, we are alluding to the complexity of the brain; in the future, perhaps, even this stage will be explainable scientifically. And from all this it follows that ambiguity must be avoided: "If the term 'interpretation', for example, means 'interpretatiom' in some cases and 'decoding' in other cases, we would have a single model that applies to everything […] Only this pre-requisite guarantees that they describe types of semiosis that actually exist in Nature. And only this can convince biologists that biosemiotics is a science, not a philosophy.[11] […] This, on the other hand, is also the precondition for a scientific biosemiotics, which tells us that building a scientific biosemiotics and a unified biosemiotics is one and the same thing" (Barbieri, 2009: 239–240). This resulted in a stormy controversy with semioticians at the pages of *The American Journal of Semiotics* (24/1–3, 2008; 25/1–2, 2009). Barbieri's main opponent was John Deely, whose model of semiotics is, in our view, crucial for understanding the semiotic nature of *all* life; however, only after our generalization, since—as we have seen above—Deely himself did not promote biosemiotics.

3.4 Troubles with the Representation

In Peircean terms, representation means that one object ("thing" in Deely) is represented by another (vehicle, representamen), whereby our experience of the vehicle allows us to somehow know the thing (an "object" in the narrower sense; Deely, 2009a: 8–13). In textbook definitions of the sign, the vehicle (*representamen*) tends to be characterized as a sense-perceptible entity that is separate from the "thing", an entity of a material or spiritual nature it represents (Krois, 2004: 106–118). But everything is more complex. For if a statue (the vehicle) represents a god, saint or demon (the object), then the actor from whose perspective this relation of representation holds does not operate in his mind with this statue, but with some *mental representation* of it, in relation to which this statue is an "object". Anything that a certain *interpretant* relates to some object, can become a sign. The process can then be repeated ad infinitum (Peirce, 1998: 69). We can also consider the "historical representation" wherein a text or a narrative interpretation is contained (Ankersmit, 1994) as metaphorically representing a certain historical period (e.g., the Renaissance). In the mind of the reader, however, the representative vehicle is not the text itself, but some *mental image*, a result of his or her imaginative fantasy, which for a particular reader (as an *interpreter*) in this situation becomes a *representamen* of the "Renaissance period" (i.e., the represented object).

Seen in this way, "signs" or "symbolic representations" can be seen with Cassirer (2020) as *internal*, mental *images* of both external objects and internal, mental

[11] Barbieri seems to assume here that there exists an absolute, let us say ontological, difference between "science" and "philosophy". From the point of view of the historical sciences, however, we can say that there is a rather smooth transition between the two, and that the difference between them is primarily methodological or epistemological.

qualities or states. Imagine a certain domain of indifferent and chaotic feelings or sensations that we experience in our innerness. Within it, our minds—in Cassirer's terms—use symbols to distinguish a *particular class* of these, into which the mind includes *precisely* certain sensations or feelings. In doing so, it not only selects them, but also brings into them a specific "spiritual energy", a specific, more general meaning content. In the process of (culturally modelled) symbolic formation, the "given" only "becomes given" (Cassirer, 1955).

Hypothetically, then, we could say that people created symbolic mental images and corresponding linguistic symbols, by means of which they could, for example, control what might appear as voices heard in the spirit. By means of these mental images, it is possible to "quieten" these voices, to release their urgency.

3.5 Semiotics and Historical Sciences

Semiotics—and especially its use within biosemiotics—strengthens the possibility to see analogies between the evolution of life the evolution of human cultures. This offers an opportunity to make some remarks on Deely's and Eco's semiotic theory of the sign also from a historical perspective; both are based on the teachings of Charles Peirce. We will be concerned primarily with the possibility of using these theories in the analysis of historical-scientific knowledge/ understanding, then in analyzing the situation of historical actors, and finally in trying to point out that presumably (almost) every process of semiosis can (temporarily) result in one form of idolatry or another (in the sense of Barfield, 1988).

In particular, we can build on Deely's distinction between things and objects (e.g., Deely, 2009a). We have seen above that objects are established in the process of semiosis in the constitution of the sign, as one of its three components (Fig. 3.1). At the same time, Deely assumes that any form of being that is not an object can become such under convenient circumstances. He notes that in the process of semiosis, both things that are in the outside world (e.g. the sun) and things that are products of the innerness (e.g., the unicorn) can become objects. At the same time, however, he rejects that the processes of semiosis are purely arbitrary. The arbitrariness of a sign is its weakness and not its strength, unless this arbitrariness is attended by the use of reason.

Deely is thus addressing a problem that is nowadays raised within the historical, cultural or social sciences by asking whether (in the socio–cultural field, i.e. realms of Closure 2 in this text) there is such a thing as *prediscursive structuration*. This question is answered in the negative by at least some of those who would attribute an immediate performative role to language or discursive practices. Others, however—for example Koselleck or Lorenz—consider linguistic or semiotic processes as formative acts, which are in one way or another commensurable, or at least potentially commensurable, with elements of extra-linguistic and prediscursive reality. In short, the processes of semiosis can also become *extra-linguistic and prediscursive* elements *of reality*: the linguistic and discursive nature of the actors can become an "object" of the historians' research, considered as independent of the language of the researchers. The elements of reality, however, are not, from Koselleck's point of view, *given* in their nature but are only *what* may show *resistance to* our interpretative ideas. This source of resistance is at least close to Deely's "thing".

One more model. The relation between "thing" and "object" (sensu Deely) or between dynamic object and immediate object (sensu Eco) has its parallel in Barfield, for example, in the form of the relation between "particles" (i.e. "the unrepresented") and "(collective) representations". "On the assumption that the world whose existence is independent of our sensation and perception consists solely of 'particles', two operations are necessary [...] in order to produce the familiar world we know. First, the sense-organs must be related to the particles in such a way as to give rise to sensations; and secondly, those mere sensations must be combined and constructed by the percipient mind into the recognizable and nameable objects we call 'things'. It is this work of construction which will here be called *figuration*." (Barfield, 1988: 24). We can infer that in his conception "particles" or a particular grouping of them, have a certain capacity to resist precisely certain historical forms of "collective representations".

Similarly, Weber (2011) judges that any part of empirical social reality can only become a process of a certain quality (i.e., actually an object of our research) if it is colored by our culturally conditioned research interest. The configurations of these parts or the factual contexts of the objects (*Dinge*) are beyond the reach of our (immediate) knowledge. We can only use ideal types to find relations between researchable problems. That we can achieve "objectivity" in our knowledge does not mean, in Weber's conception, that we recognize the ontic connections of parts of social reality as they really are.

Seen from the noetic perspective of historical sciences, we can say that Barfield's "particles" and their possible but unknown grouping, just as Weber's "parts" and their possible but unknown grouping, are parallel to Deely's "things". The capacity of these "things" to do so is nothing more than the capacity to resist, to *create resistance to our interpretative ideas* (in line with Koselleck's thesis on the "theory of possible history" and the "veto of sources"; see 5.2). In this sense, our theoretical or narrative interpretations are not arbitrary. According to Iggers and Chartier, the "constitutive element of historiography" is the relation to a reality that preceded the historical text and existed outside it; the historical text, then, is supposed to give an intelligible account of what happened (Iggers, 1997: 20–21; Chartier, 1987).

We could then say, in Deely's terminology, that the historical text, including the theories and narrative interpretations contained therein, deals with "objects", whereas the "reality" that exists before and outside of the text would be a chaos of "things". Theorists of history will differ as to whether, and if so to what extent, we can use (theoretical and narrative) mental operations on "objects" (facts, events, processes, structures, relations, etc.) to penetrate to the true nature of "things" and to discern the real connections between them. Those who share the ontological paradigm would usually answer this question in the affirmative. From the perspective of the mentalist paradigm, it would probably be generally assumed that we are able to model "objects" and the correlations between them. We can judge the adequacy of both models by confronting them with source traces and achieve a certain "objectivity", yet the true nature of "things" remains elusive for us. For many adherents of the linguistic paradigm, the relation to a reality that would exist before and outside a text is inconceivable.

However, the matter is not so clear-cut. Is there indeed a reality that precedes the actor's text and exists outside it? This question could perhaps still be answered in the affirmative. But is there a "reality" that precedes and exists outside the actor's linguistic acts or acts of semiosis?[12] This question can no longer be answered in the affirmative without further explanation, if it can be answered in the affirmative at all. The integrating role of narrative sentences can be transferred from historical interpretations to the actor's experience of lived situations. The content of narrative sentences is certainly more often controlled by rational historical-scientific methods than the content of the narrative sentences of actors. It could be said, then, that the narrative sentences of actors will tend much more towards arbitrariness than the narrative sentences of historians. But one might object: arbitrariness in respect to what? After all, actors also directly *co-create* their reality with narrative sentences. The core of the problem, however, lies in this "co–creation". If there are processes of semiosis taking place in the individual minds of the actors—or even in a certain transcendent field of collective mentalities, shared experience and memory—then there is also a tension between the objects established in the formation of the respective sign on the one hand and the things on the other. The results of the processes of semiosis or the contents of narrative interpretations can—sometimes immediately, sometimes after a long time, turn out to be wrong or become dysfunctional in the given socio–cultural field. Examples include narratives that constitute national identity.

3.5.1 Arbitrariness

Hence, the story, narration, narratives that constitute a particular community need not be a construct and has an essential, ancient existence. The solution might perhaps be found in the fact that "thing" like the "nation" is indeed a cultural construct, but not a voluntarist invention. The project of the "nation", formulated by intellectuals, is subject to potential resistance of a given socio–cultural field. It depends on the chance that some subjective meaning of action will achieve its objectification, i.e., will be *understood and accepted*, or tolerated, in a particular field. Not every subjectively meant sense that is publicly articulable (text, speech) has an (equal) *chance of* reaching its *objectification*. At the same time, if it reaches it, it does not mean that other sense, even contradictory, cannot be objectified in the same field (e.g. "the idea of the Czech nation" and "the idea of the German nation" in Bohemia).

[12] These are the questions like "Is there a moon when no one is looking?", or "Were there bacteria before Pasteur?". They appear with iron–clad regularity in discussions at seminars on the subject.

The polysemy of features or the varying degree of chance of a particular process of semiosis or a particular narrative to achieve its objectification can be *acknowledged* in principle, but it is difficult to *measure* it in concrete cases. The history of human culture, seen in a macroanalytic perspective, provides many examples of competing cultural variants of a particular invariant that—one can say with varying "chances"—have maintained themselves in one sociocultural field. In such cases one can say which one was more successful in that field than the one competing with it.

An example of this is the advocacy of baptism in adulthood among the Anabaptists; this was a socio–cultural phenomenon in central Europe in the sixteenth and early seventeenth centuries. It was *comprehensible*, *objectified*, but conspicuously a *minority* concerning interpretation of the sacrament. However, to claim that this chance can be measured by the percentage of adherents of various denominations would be an inappropriate simplification. One could also find cultural variants that did not reach their chance: for example, the federative multi-ethnic organization of the Austrian Empire in 1918.

The socio–cultural field (i.e. the space in which Closures 2 can be grouped) is capable of holding and preserving for a shorter or longer time even very peculiar interpretations (i.e. different Closures 2). For example, if one insists that the "texture" of the way the man I just met crosses the street is an expression of his ethnicity. Or if I'm a dogmatic Marxist and insist that the Anabaptists adherence to baptism in adulthood is an articulation of the economic status of the poor. Seen from this perspective, signs or narratives may seem arbitrary. But they are not arbitrary, at least in the sense that they are marginalized or suppressed and even refuted. This can happen in very different ways. In a very rational way, for example if the sciences are subjected to systematic methodological critique, or in an emotive way, when certain narratives do not suit the "mood of the times". The linguistic or semiotic arbitrariness (of which, as mentioned above, Deely speaks) need not be and is not an arbitrariness in terms of the processes of semiosis taking place in minds, in terms of noetics, or in terms of the sociology of knowledge.

Moreover, the history of science and the history of human culture often show directly opposite tendencies. Certain interpretations begin to be attributed a directly "thing-like" nature (in the sense of Deely's notion of "thing"). This undeniable "thingness" has been attributed by historiography, for example, to "Old Slavic equality and democracy" or to the idea of "slavery on the large estate" (see Chap. 7). The interpretation of sacraments in cultural history has repeatedly turned into idolatry (see below). Such reified "objects" can, and often do, come into tension with "things". The result, then, can be what Lotman (2009, chapter 4) refers to as a "cultural explosion" or the fact that, as would say, the spontaneity of "life" always breaks through established "cultural forms" after a time.

If we are discussing here the analogies of (the development of) biological species and (human) cultures, it would not be without interest to see to what extent we can determine the chance of (bio)semiotic processes to achieve their objectification, or to what extent we can identify unsuccessful attempts at the level of life to interpret the records.

References

Ankersmit, F. R. (1994). Six theses on narrativist philosophy of history. In *History and tropology. The rise and fall of metaphor* (pp. 33–43). University of California Press.

Augustine, A. (1997). *[426] On Christian teaching*. Oxford University Press.

Barbieri, M. (2003). *The organic codes. An introduction to semantic biology*. Cambridge University Press.

Barbieri, M. (2009). A short history of biosemiotics. *Biosemiotics, 2*, 221–245.

Barbieri, M. (2015). *Code biology. A new science of life*. Springer.

Barbieri, M. (2024). *Codes and Evolution: The Origin of Absolute Novelties*. Springer.

Barfield, O. (1988). *[1957] Saving the appearances. A study in idolatry*. Wesleyan University Press.

Cassirer, E. (1946). *The myth of the state*. Yale University Press.

Cassirer, E. (1955). *The philosophy of symbolic forms. Vol. 2: Mythical thought*. Yale University Press.

Cassirer, E. (2020). *The philosophy of symbolic forms. Vol. 1: Language*. Taylor & Francis.

Chartier, R. (1987). L'histoire ou le récit véridique. In *Philosophie et histoire* (pp. 115–135). Centre Georges Pompidou.

Deely, J. (2008). *Descartes & Poinsot. The crossroad of signs and ideas*. University of Scranton Press.

Deely, J. (2009a). *Purely objective reality*. Mouton de Gruyter.

Deely, J. (2009b). *Augustine & Poinsot. The protosemiotic development*. University of Scranton Press.

Deely, J. (2010). *Medieval philosophy redefined Scranton*. University of Scranton Press.

Eco, U. (1999). *Kant and the platypus. Essays on language and cognition*. Harcourt Brace & Co.

Heidegger, M. (1995). *[1929/30] The fundamental concepts of metaphysics*. Indiana University Press.

Iggers, G. G. (1997). Historisches Denken im 19. Jahrhundert. Überlegungen zu einer Synthese. In W. Küttler et al. (Eds.), *Geschichtsdiskurs, Band 3: Die Epoche der Historisierung* (pp. 459–470). Fischer.

Krois, J. M. (2004). Kultur als Zeichensystem. In F. Jaeger & B. Liebsch (Eds.), *Handbuch der Kulturwissenschaft, Band 1: Grundlagen und Schlüsselbegriffe* (pp. 106–118). J. B. Metzler.

Lotman, Y. M. (2009). *[1992] Culture and explosion*. Mouton de Gruyter.

Markoš, A. (2002). *Readers of the book of life. Contextualizing developmental evolutionar biology*. Oxford University Press.

Peirce, C. S. (1998). [1903] Nomenclature and divisions of triadic relations, as far as they are determined. In E. Peirce (Ed.), *Peirce selected philosophical writings 2 (1893 1913)*. Indiana University Press.

Poinsot, J. (2013). *[1632] Tractatus de signis. The semiotic of John Poinsot* (2nd ed.. Latin–English bilingua; selected, translated and commented by J. Deely). St. Augustine Press.

Ricoeur, P. (1984). *Time and narrative I*. University of Chicago Press.

Weber, M. (1904). Die "Objektivität" sozialwissenschaftlicher und sozialpolitischer Erkenntnis. *Arch Sozialwissenschaft und Sozialpolitik, 19*, 22–87.

Weber, M. (2011). *Methodology of the social sciences*. Routledge.

Chapter 4
Memory and Explosion

Abstract An individual as a connective structure: an individual has memory, but it is always created collectively. An individual is always part of what we call Closure 2, is member of many interconnected groups; a multicellular individual as our body is itself a closure 2. The present as a frame, as a reference for remembering—or re–interpreting, even of an invention of the past. Mimetization or reconfiguration of values: Lotmanian explosion. Language is not a simple, unchanging code, but a code in a history; the past is a cultural creation. Communicative and collective memory, with a floating gap in between. Translations of meaning—between languages, or between historical forms of the same entity. Distributed language as a pool of nonwritten memory. Forms of written messages, texts, reading, understanding. Model and empirical author and/or reader: how understanding will appear; abductive procedures and presuppositions; small worlds. Alphabetical and pictogram recordings of speech.

Keywords Memory as a collective structure · Singularities (*explosions*) · Language and code · Communicative and collective memory · Speech and script

> Roughly speaking, physics has ended by having to conclude that [...] the whole world of nature as we actually experience it depends for its configuration on the mind and senses of man. It is what it is because we are what we are. Thus our common assumption that the main effort of human thinking has been to make a mental replica of a pre-existent outer world is incompatible even with the scientific approach to things out of which it arose. This assumption is indeed determined by science; but by a science of the day before yesterday.
>
> Owen Barfield, 2013: 22.

The title of this chapter is a "recombination" of two titles of books from the 1990s. The historian Jan Assmann writes *Cultural Memory and Early Civilization* (2011), and the semiotician Yuri Lotman writes *Culture and Explosion* (2009). The reader already knows that we are trying to extend the term "culture" to all communities of

A. Markoš, J. Horský, *The Seeds of Forms and the Forms of Seeds*, Biosemiotics 12, https://doi.org/10.1007/978-3-032-16131-4_4

living creatures: every Closure 2 is a culture. The authors mentioned above do *not* attempt to do so, of course; their concern is with the cultures of societies exclusively of human beings, yet they are a tremendous inspiration to us. Assmann states right at the beginning of his book:

> Every culture formulates something that might be called a *connective structure*. It has a binding effect that works on two levels—social and temporal. It binds people together by providing a 'symbolic universe'—a common area of experience, expectation, and action whose connecting force provides them with trust and with orientation. [...] However, it also links yesterday with today by giving form and presence to influential experiences and memories, incorporating images and tales from another time into the background of the onward moving present, and bringing with it hope and continuity. This connective structure is the aspect of culture that underlies myths and histories. Both the normative and the narrative elements of these—mixing instruction with storytelling—create a basis of belonging, of identity, so that the individual can then talk of 'we'. What binds him to this plural is the connective structure of common knowledge and characteristics—first through adherence to the same laws and values, and second through the memory of a shared past (Assmann, 2011: 2–3; italics author).

And another very important statement: "Despite the fact that it is always the individual who 'has'memory, it is created collectively. This is why the term collective memory should not be read as a metaphor, because while the group itself does 'have' a memory, it determines the memory of its members. Even the most personal recollections only come about through communication and social interaction. We recall not only what we have learned and heard from others but also how others respond to what we consider to be significant" (Assmann, 2011: 22).

The individual is part of Closures 2 that shape, co-define him or her. And for a given culture (group): "Whoever shares it thereby demonstrates his membership in the group, and so it is not only bound to time and place but also to a specific identity. In other words, it is related exclusively to the standpoint of one real and living community. In the context of the above, we add: not of a group, but of groups" (Assmann, 2011: 25). Statements above /and also what follows) could be—in our opinion—easily adopted to relationship between an individual and its species, or between species and the community of its ecosystem.

The cohabited past cannot completely disappear yet must be distinct from the present (the present forms a frame of reference for remembering), but let us add that it can also be "invented"—for example, on the basis of the mimetization of another culture, or on the basis of a total reconfiguration of values—this is the "explosion" in the title of Lotman's work. The two authors mentioned (and with them many others), emphasize precisely the dynamics of culture, even in case of a culture with writing, which rather seems to seek preservation, stability of knowledge. An absolute preservation, however, would mean that the whole of history (and likewise the evolution of life) would be turned into a mere chronology and the story would disappear from it entirely. Recall, for example, the constant, centuries-old interpretation of the Bible—both linguistically (translations) and simply historically given. Nothing is eternal and ideal; trends towards stability even in cultures long established are constantly being fluffed up—not just through interpretations of texts, but through changes in perceptions of the world in general. Moreover, there are also frequent returns (even in the form of "explosions") to tradition, or rather "tradition", i.e. to a state of affairs sometimes reinterpreted on the basis of sources, sometimes simply whitewashed or inadequate.

> The moment of explosion is also the place where a sharp increase in the informativity of the entire system takes place. The developmental curve jumps, here, to a completely new, unpredictable and much more complex path. The dominant element, which appears as a result of the explosion and which determines future movement, *can come from any element of the system or may even be an element of another system*, randomly pulled by the explosion into the web of possibilities of future movement. [...] By removing the moment of unpredictability from the historical process, we make it totally redundant. From the position of a bearer of Reason who holds to the process of the internal point of view (be it God, Hegel or any other philosopher, who has mastered the 'singular scientific method'), this movement is deprived of informativity (Lotman, 2009: 14; italics ours).

Lotman criticizes attempts to reduce natural language and its texts to (unambiguous) code—as if the task of the semiotic process of communication were to transmit messages: "The abstract model of communication not only implies the use of one and the same code, but also the identical memory capacity of both addresser and addressee. In fact, the substitution of the term 'language' by the term 'code' is not as harmless as it seems. The term 'code' carries with it the idea of an artificial, newly created structure, introduced by instantaneous agreement. A code does not imply history, that is, psychologically it orients us towards artificial language, which is also, in general, assumed to be an ideal model of language. 'Language', albeit unconsciously, awakes in us an image of the historical reach of existence. Language is a code plus its history" (Lotman, 2009: 4). (Recall at this point M. Barbieri's organic codes from the previous chapter.)

It is up to the recipient to accept the text as code or as a message in natural language; moreover, communication between different organizational or temporal domains can never be canonical, with predefined rules. Above all, it is a question of "its relation to the extra-system, to the world which lies beyond its borders (...) At the same time, the question arises as to the level of adaptability of a world created by language towards a world existing outside of language, i.e. beyond its borders" (Lotman, 2009: 1). "Thus, the world of semiosis is not fatally locked in on itself: it forms a complex structure, which always 'plays' with the space external to it, first drawing it into itself, then throwing into it those elements of its own which have already been used and which have lost their semiotic activity" (Lotman 2009, 24). Everything said so far in this chapter can, as we intend to show, be extrapolated to living beings and their communities: semiosis extended to biosemiosis. And there is one more observation that also follows for all living things: Assmann (2011: 33) refers to Halbwachs (1980), who apparently was the first to point out at many places that the past is not a natural growth but a cultural creation.

4.1 Where Memory Ends, History Begins

The play between norm and lived presence, so important for both kinds of closures, we focus here especially to Closure 2, i.e. for communities of all types. (For Closure 1, see Markoš & Švorcová, 2019, for details). Assmann adopts Vansina's (1985) division of the cultural memory of a community into *communicative* (generational) and *collective*, with a *floating gap* yawning between the two. The division does not depend on whether the cultures are oral or scriptural. If we talk about human cultures, *communicative memory* is formed from the experiences of the currently living members of the community and what they may have learned from their immediate ancestors, i.e. it spans about 40–80 years. Amnesia sets in about earlier events—except for what has been stored in the *collective memory*, which represents the tradition and founding myths that bind the whole community together (deeply rooted norms kept by the community; Rappaport, 2010). Myths transcend the mundane, they are "over life-size", providing filters through which even older general norms of community life are viewed. Both forms of memory interact with each other and "create" (invent) the past to *suit* the community's self-understanding and its position in the totality of all other communities—especially those in immediate contact. Perhaps all the following chapters in our book provide an illustration of this arrangement.

If we add to this the written culture (see also *hypolepsis* in 2.3), the community experiences yet another kind of evolution: "Some texts disappear, some enter the flood, and others are extended, shortened, revised, or anthologized in ever changing combinations […] Some assume central importance because they are copied and quoted more often than others; thus they eventually become classics that embody normative and formative values. […] In this way a 'great tradition' emerged that created a repository of knowledge and, more importantly, education that extended back over hundreds and then thousands of years. […] Canonical texts cannot be changed—this marks the crucial difference between them and the stream of tradition. They are sacrosanct, and must be handed down word for word" (Assmann, 2011: 76–78).

There is also the opposite current—the necessity of interpretation (and often translation into another language) of such frozen canonical texts "There was always a gap between fixed text and a changing reality because the letter could not be changed one iota, but the human world was constantly subject to change; this gap could only be bridged by means of interpretation. The latter therefore became the central principle of cultural continuity and identity" (Assmann, 2011: 80).

Culture becomes "elastic" (analogous to model of Flegr, 2010; see Chap. 10) and absorbs (resists?) a series of fluctuations with the help of such a structure. Those parts of the canon that are not interpreted become a "grave of meaning" and fall into oblivion. There is, of course, a way out: since not a single comma on the canon may change, it is still available at all times and its elements can re-emerge into a living presence when the need arises. This allows for "plastic" phases of culture, where such resurrection restructures its overall condition: "Every period has its own canon.

Such shifts are only possible when that which has been excluded remains dormant in the cultural memory and does not disappear under the weight of total condemnation" (Assmann, 2011: 103). "Times of internal cultural polarization, when traditions are broken and people must decide which order to follow, result in the formation of canons. In such situations, the canon embodies each competing order, and each canon lays claim to being the best or indeed the only true tradition. Whoever subscribes to it, automatically subscribes to a normative definition of oneself or to an 'identity' that will be in harmony with the demands of reason or of revelation" (Assmann, 2011: 106–107). And there are still plenty of texts that are not canonical; it is possible (and necessary) to rearrange them in all sorts of ways into new contexts and shape the narrative accordingly—all of which is a field both for the historian and the evolutionary biologist.

As we have set out, we also look for analogies in non-human communities—this is the subject of Chaps. 9, 10 and 11 in particular, where we will discuss transgenerational epigenetics and the normative "tradition" of life lineages, as well as the plastic and elastic phase of the evolution of such lineages.

4.2 Interpretation of the World

Barfield's epigraph for this chapter seems almost like an invitation to a Postmodern *avant la lettre*. What is it then? Is the world always different, as are also we at any given moment? Yes and no: as Barfield constantly points out to us in his writings, one's *participation* in the world is constantly changing with the times and the community in which one lives—and, of course, with language. And knowledge in these domains is, as we have already noted, the concern of any number of "humanities". If we are committed to proving the semiotic nature of, we must first consider language and then semiotics in particular. We will then try to bounce into biosemiotics and argue that *every* living creature is capable of interpreting the world in this way, and that *this* interpretation also changes with time and community—in other words, it is subject to evolution (or resistant to it) just as much as the human human evolution is. Thus, we come up against the ever–recurring question of whether or not the interpretation of the world is dependent on thought and articulated speech—that is, whether *every* living thing is really capable of semiosis, or only just humans are.

In this section we shall dwell on the development of language and the development of the meanings of words as well as the difficulties of translation, and then we must necessarily deal with written and spoken forms of language—their mutual reduction and development. We set the mood by quoting Stefan Zweig, who says: "Just because every strange language at first offers opposition in its most personal turnings to those who would copy it, it invites forces of expression which, *otherwise unsought, would never come to light;* and this struggle to wrest from a strange language its most intimate essence and to mold it as plastically into one's own language, was always a particular artistic desire on my part" (Zweig, 1942: 92; italics ours).

Let's start with a "verbatim" translation between languages—Darwin's terminology may serve as an example. The Czech translation of *fitness* [zdatnost] is also how Czech philosophers translate the ancient Greek word *arété*. However, the English translation of the word *arété* is not *fitness*, but *excellence*! We feel the affinity, but is it the same? Similarly, we will be at a loss to translate the word *struggle*;[1] and a perennial problem concerns the translation (into Czech) of Darwin's notion of *descent with modification*. It is similar with the translations of the terminology of philosophers. We mention similar problems elsewhere; for example, physical terms taken from natural language: they are wrapped in clouds of meanings that blur the precise meaning of the term. For example, does *today's* meaning field of the English word *force* coincide with the same sounding word of Newton's time? Is it spelled the same, is it even pronounced the same? And if it is pronounced differently, hasn't pronunciation moved it into the neighborhood of other words, into different contexts? The translation of such terms can also involve interesting shifts; see below (4.7) on the Chinese equivalent of "entropy". Context also plays an important role—for example, see also below the anecdotical term "mondegreen", or perhaps a "new" interpretation of Wordsworth's verse "a poet could not but be gay". In Čapek's story (2.1) do the archivist and the inspector interpret the texts from fifteenth century correctly—including the necessary reading between the lines?

We judge that even homonymous forms (of proteins, protein domains, cells, attitudes and characters) play a similar role in expressing the Appearances of the living, as well as in their *recognition* of other living beings: this includes, for example, symbioses that are not merely sharing metabolites. Or: what predisposes some species to become invasive? A game of synonyms, metaphors, analogies and parables everywhere.

But back to the language. The historical development of language poses many questions for literary scholars, historians, historical linguists and others. Today's native speakers of English certainly have trouble reading and listening to Shakespeare—yet, as we read somewhere, it caused a storm when it was recomposed into contemporary English—for school children. Yet no one disputes the words of Trask (2010: 156): "Shakespeare's written English already seems very strange to us, and we probably wouldn't understand his speech at all if we could hear it." So, what do we receive by watching Shakespeare in the theatre today? A reconstruction of his world? A construction, i.e., a re-creation? Certainly, it does not hide its descent, but with how many new modifications? We also refer to the following chapter, where, among other things, the re-edition of Czech texts, long after their production, is discussed. And what about the texts of, for example, Renaissance mystics and alchemists, texts written in different dialects: how to decipher their idiosyncratic terminology? The further we go into the past, the more difficult things become: how to trust, for example, a translation from Sumerian, the Indian Vedas, or the Book of the I Ching?

---

[1] It is translated rather as "fight" in most, if not all, languages of the Continent. An English colleague of ours was quite surprised: well, *struggle* is more like *striving*, he said. We feel the difference—but does it or not fundamentally affect the interpretation of the context in translations?

4.3 Writing, Speech, and Distributed Language

In Italo Calvino (1981: 139) we read: "It is only through the confining act of writing that the immensity of the nonwritten becomes legible, that is, through the uncertainties of spelling, the occasional lapses, oversights, unchecked leaps of the word and the pen. Otherwise, what is outside of us should not insist on communicating through the word, spoken or written: let it send its messages by other paths." Take the relationship between verbal expressions and their notation: *from the point of view of each, the other represents a considerable reduction*—and at the same time it offers an opportunity for mutual enrichment, as well as an opening for new ways of understanding the world. Our task, then, is to evaluate the relationship of speech notation to speech itself—but also, conversely, the influence of speech on the mode of notation. And since our inquiry concerns all of life, it will also be about expressive media other than speech and its notation—gestures, attitudes, structures, signals and signs, in a certain context. Gestures, facial expressions, voice modulation can certainly do this: just think of what can be expressed with an upraised finger—first, second, or third! Immediately, also genetic script and its interpretation strike the mind.

At this point, we touch the doublet Phenotype—Appearance, that will be discussed throughout our text (especially in Chap. 8): for now, only a brief example. Let us compare the differences between the Appearance and phenotype of the tern *Sterna paradisaea.* This bird of the polar regions flies every year between the Arctic and Antarctica, always to be at the site of the polar summer. Its Appearance is associated with countless *phenotypic* traits such as size, weight, flight endurance, plumage structure, nutrition, anatomy, nesting, etc. Indeed, a stuffed exhibit in a display case or a picture in a textbook is only a faint echo of the Appearance of this magnificent bird. A biologist must necessarily simplify: so many biologists use "phenotype" to mean "Appearance"! Instead of a tern, consider, say, a class of pupils—what all researchers must have in order to get reliable *phenotypic* data of 12-year-old boys?

The topic was developed at the beginning of this century by linguists associated under the name of the *Distributed Language Group.*[2] The authors point out that we *enter* our language, we do not learn it, it is an *extension* over our communicative and cognitive processes. On one hand the extension involves bodily functions, on the other interconnections between society members as a process of "languaging". It is thus a first-order communicative activity, from which linguistic utterances emerge as a kind of reduction. Because language can be analyzed and described (spelling, grammar, etc.), and because there exists (in our tradition) yet another reduction in the form of alphabetic writing, the idea has taken hold that speech and written form

[2] See, for example, the special fifth issue of *Language Sciences* 29, 2007 and other articles in this and other journals. They have also later been identified under the heading *Cognition beyond the Brain*; for example, Cowley, 2011; Cowley & Valée-Tourangeau, 2013; Harris, 2009. See also Markoš et al., 2013; Švorcová et al., 2018.

are digital products of rationality (of the mind), and moreover that they are "hard-coded" like the Morse code; bodily functions and their interconnections are usually not considered (e.g. Love, 2004, 2007).

But language and its notation are only two of the many forms of a rich communicative game—and only between humans. Is it possible to remove language and writing and still talk about forms of culture, say, in a bacterial coating or a meadow? Are we in such cases confronted with the "speech" of interacting collectives whose members *understand* each other? If we stay with human culture first and try to illuminate the nuances of the relationship between speech and writing by comparing the two most widely used forms of writing today: the "Latin" writing with the alphabet[3] and the "Chinese" (or better, East Asian) writing with the help of special characters for the morphemes of the language. This will then provide us with analogies to compare genetic notation and the *Appearance* of a living creature.

Let's start with a well-known phenomenon which goes by the technical name of *mondegreen*. It originated when the verse "laid him on the green"[4] (perhaps "laid him on the meadow", though without context it could also be understood as "laid him on the ping-pong table") was taken for "Lady Mondegreen". Such shifts occur frequently, often occurring as part of jokes, but also as part of serious-faced nonsense. Note that these are mostly verbal expressions, caused either by a slip of the tongue, a linguistic slip, or a (often deliberate) misunderstanding of the term. This is akin to swapping syllables (puns) or letters (anagrams) in words. Alphabetic writing will quite readily reveal typos and nonsense—if, of course, what is said is well heard or, in case of a script, we know grammar and spelling—otherwise further mutations may occur.

We know that we think, we know that we need the brain to do so, and we know a lot about how it works, we know which structures are indispensable and what happens when they are damaged (whether genetically, by injury or pharmacologically). But how this or that (physiological, structural, cellular, etc.) process *contributes* to the fact that we *somehow* think and act on the world, we do not know. It is up to the reader to add "yet", or to believe that while we all have neurons, electrical impulses, etc., how consciousness grows out of this seedbed is an individual matter, unique to each of us, and epigenetic to a large, perhaps decisive, degree. Similarly, uniqueness may also manifest itself at the level of cellular events in ontogeny or in evolutionary lineages. Jellyfish do not have brains, yet one group (cubomedusae) is equipped with 24 eyes, 8 of which are homologous in structure and genetics to the human eye. What does a jellyfish see, and how does it perceive its world on that basis? (O'Connor et al., 2009; Kozmik et al., 2008).

On the one hand, life has encoded—at different levels—often ancient semantic codes (grammars), on the other hand, it uses them to express subtle variations,

[3] The alphabet is of course also used by Semitic languages and many others, but here we will limit ourselves to the Indo-European Language Family. Even so, this poses many problems, starting with the transliteration between the Latin and Greek alphabets and their many derivatives.

[4] From the verse of a Scottish ballade "Oh, they have slain the Earl O'Morray and laid him on the green" misheard as "Oh, they have slain the Earl O'Morray and Lady Mondegreen."

interpretations of such canonical "texts" and practices, all in the context of external conditions. In each of these cases, a vast number of local and temporal actions coalesce into a kind of dynamic, holistic "field" that encompasses momentary contexts, confronting them with memory and experience, and also with the momentary "horizon of expectation" (sensu Koselleck, 2004).

4.4 Text and Reading

We are alive, and our culture and technology are therefore products of life. We (the authors) therefore find it fruitful to look at some aspects of life in an anthropocentric way, as opposed to the more common view from the point of view of molecules, i.e., the way of explaining life from the inanimate. Of course, the anthropocentric approach has always been at work here too, it is just not felt as such and when recognized it is rejected. But what else is "mechanistic natural science" but a constant search for analogies to the products of our culture, be they machines of the first, second, or however many steps of the industrial revolution: initially levers, pulleys and wheels, later heat machines and organic chemistry, then transcription, translation, code, and still later computers and artificial intelligence… We therefore consider it important to dwell on the possibilities offered by cultural sciences: their experience in interpreting music, art and texts. Perhaps this will help us understand how a living being might interpret itself and its biospheric commonwealth. We will follow the dialectic of two opposing pressures—on the one hand, the norms, manners, canons and even taboos established in the past, which should have to be respected, and on the other hand, the freedom of the interpretative process.

We leave aside texts such as the law codices, manuals, warning inscriptions or road signs, where care is taken to *minimize* the field of interpretation—although even here, as we know, it is not an entirely machine-like language: even a canon implicitly refers beyond itself. For example, in the Middle Ages, the topic of a painting was not depicted, yet it remained implicitly present for those who understood biblical or other contexts; the image served as reminder, whether by imitation or guidance. But then the onlooker must also be given some kind of external knowledge: "The image—in which we often see something other than what is actually shown—can lie in a much more subtle way than written or spoken language. If we want to keep our visual memory in any way intact, we must teach the coming generations how to look at images" (J.-C. Carrière, in Carrière & Eco, 2012: 159). Moreover, "[The medieval interpreter] behaves as if there was only one code, whereas everyone knows that there are many, but nobody seems to mind. In this way, if we must speak of a medieval symbolic code, we must bear in mind that it is a code full of semantic bifurcations, an honest-to-goodness dictionary of synonyms and homonyms, in which one image may suggest many realities and a real object many different images—all of them true because what God has to say is vast and complex, every language inadequate, and we must try to grasp it as best we can, a little at a time" (Eco, 2014: 259).

The interpretative approaches of the one who creates and the one who perceives can also be quite diverse. If we restrict ourselves to literary works, a quartet comes to mind—the model author, model reader, empirical author and empirical reader, and these modes are discussed by Umberto Eco in many of his works (e.g. Eco, 1997; see also Calvino, 1981; see 1.2).

We assume a model author in the text even before we start reading. Needless to say, the model author differs from the empirical author, the actual person who wrote the text (with his/her life peripeties, etc.). Moreover, few people read a work immediately after it has appeared—and any delay modifies the interpretation—even for the author himself, should he or she still live. The readers themselves should nevertheless seek rules that allow them to remove ambiguities: the text is a tool for the author to produce a model reader (anticipating his mental "phenotype")—and the reader in turn tries to discover in a similar way the model author, as we read elsewhere (Eco, 1990). There is a long–term mutual cultivation of author and reader—see the illustration of the "co–cultivation" of God and His people as given by Fromm (2013).

A kind of ideal *model reader* is formed in the imagination of the author; such a reader will interpret the text exactly as the author "meant" it. However, the "the model reader of a story is not the empirical reader. The empirical reader is you, me, anyone, when we read a text. Empirical readers can read in many ways, and there is no law that tells them how to read, because they often use the text as a container for their own passions, which may come from outside the text or which the text may arouse by chance" (Eco, 1994). The empirical reader can play the role of a model only if he or she knows the author well and *chooses* to read in accord with his or her intentions, or if he or she is indoctrinated in advance into the author's mode of understanding: hence the crucial role of teachers, clergy, politicians and other authorities in our culture. Lest it sound so trite, let us again recall the empirical readers of canonical Christian Scriptures and how they were understood in different times and different settings as well as through different translations. The whole history of Christianity is carried by the ethos of understanding—and interpreting for others—what Salvation consists of; and then disposing of the "heretics", the bearers of different interpretative schemes of the given period (e.g., Gurevich, 1990; Ginzburg, 2013). Recall also the dynamic interpretations of the history of different societies where even the text itself is subject to unexpected twists and turns.

We now invite the empirical reader of this text to try to imagine the cell as an empirical reader: the reader of its inherited, largely conservative genetic text, read in confrontation with the external context (in particular, the quality and setting in of the various Closures 2 of which it is a member). An *informed* reader, providing that too arbitrary interpretations with a tilt toward novelty will be—by the cellular community—usually suppressed. Isn't understanding *this* type *also* the driving force behind evolutionary pressure? Imagine even the cell as a creative *author* of genetic text—in particular, the prokaryotic world is quite inventive in this regard.

What does a reader who knows the code (i.e., who can read the text) actually do? He or she usually knows the background, presupposes the meanings of the text, and can usually arrive rather quickly at an interpretation of the meaning of what he or she is reading in accord with established conventions and contexts. Readers also add their own memory, experience and narrative habits to the interpretation: after all, this is why they are empirical readers, unique and individual, and not manufactured on a conveyor belt. A special case is *abduction*, the hypothesis of meaning in cases where conventional interpretations of codes cease to work. If such an abductive attempt proves successful and is accepted by the community, it becomes part of the network of meanings of the language. However, abduction is *always* present in the background, which is precisely why the interpretation of a text is not clear-cut (Eco, 1990). Abduction helps us choose an appropriate decoding system for indeterminate messages (e.g., what "OCC" means in Čapek's tale in Chap. 1; see also Ginzburg (1989) who stresses forensic facets of understanding). The creativity of interpretation has two components: creativity that preserves the rules (*norms* in Markoš & Švorcová, 2019) and creativity (game over the rules) that changes the rules themselves and allows for invention, novelty.

With genetic texts and structures, an "intentio operis" is a tricky thing. The "model author" in this case is the evolutionary process, but often these texts have long since frozen at the level of canonical codes (see Chaps. 9, 10 and 11 for some examples) and therefore require many number of modifications to suit the times. The pinnacle of the interpretative art of "empirical living beings" is the construction of species-specific Appearance based on the interpretation of very similar genetic transcripts—cf. for example, the genomes of human and chimpanzee (e.g., Carroll, 2006; Carroll et al., 2001; Gilbert, 1994). Such subtle *epigenetic interpretations* are outlined below and discussed in more detail in Chaps.10 and 11.

Readers accede to the game of a fictional text—a *small world*; they make their own assumptions about things that are not explicitly stated in the text, and then formulate their understanding of the story from their overall disposition and experience—and from keys extracted from contexts. Umberto Eco's *Open Work* (1989; see also Eco, 1994) explores this fascinating area from different angles.

An analogy is immediately offered with the reuse of genetic or regulatory modules in ontogeny and physiological function. The classic example is the module of homeotic (hox) genes: initially their sequential activation determined the anteroposterior axis of the animal; today this is also true, but over the course of evolution they have also been used (in vertebrates) to determine, for example, the order of the fingers on the hand, the order of the parts of the limbs, as well as of various organs (e.g. Cowley & Markoš, 2019).

The structure of artistic texts also reflects the time and place of their creation, and with a shift in vocabulary, and a change of implicitly perceived realities, worldviews and perception of how that time is viewed today (see also Lotman above), we can arrive at interesting new meanings of the work. However, there may also be a reduction. In 1999, there was an interesting exhibition in Oslo called "The Year 2000 through the Eyes of the Year 3000": besides rusty fenders, rotten condoms and other everyday objects from our time, there were a striking number of objects labelled "votive": the most mysterious of these were perforated brass rings with engraved numbers.[5] What we cannot explain is declared to be a votive object… When we try to interpret the forms of organisms of different times and environments, are we not also committing such a reduction?

Analogies with biological evolution provided by the book of interviews quoted above (Carrière & Eco, 2012) are also extremely interesting: "Over time, every book is overlaid with all the interpretations that have been made of it. We don't read the same Shakespeare that Shakespeare wrote. Our Shakespeare is much richer than the Shakespeare that was read at the time. A masterpiece isn't a masterpiece until it is well known and *has absorbed all the interpretations to which it has given rise*, which in turn make it what it is. An unknown masterpiece hasn't had enough readers, or readings, or interpretations" (Carrière & Eco, 2012: 102, Eco speaking, italics ours). We know many contexts of the time that Shakespeare could not have known—and because of this, our understanding of the text is richer. Well, is it? Can we recognize *all* references, jokes and puns circulating in his lifetime? The exegesis of Scripture, or of famous philosophers from different ages, or simply a history tale written in Čapek (Chap. 1) give even richer examples.

And one more observation, this time by J.-C. Carrière: "We can of course explain the great influence that Cervantes had on Kafka. But we can also […] say that Kafka has had an influence on Cervantes. If I read Kafka before reading Cervantes, then through me and without my knowing it, Kafka will impact on my reading of Quixote." (Carrière & Eco, 2012: 120; see also 2.3, hypolepsis). We evaluate the past, even the evolutionary past, through the perspective of today: "Something no doubt occurred, and what is really being propounded by such popular writers, and, so far as I am aware, by the text-books on which they rely, is this. That at that time the unrepresented was behaving in such a way that, human beings with the collective representations characteristic of the last few centuries of western civilization had been there, the things described would also have been there" (Barfield, 1988: 37; see also Merton, 1967). That the past is malleable and new ideas can be discovered in it, ideas that the "elders" themselves did not recognize in their works somehow escaped us all, yet we often dig out from the old authors what they did not say. Did Empedocles discover evolutionary thinking? Or was it Heraclites? Was the future history of the world predicted? Or, on the contrary, is there nothing new under the sun, or even before the sun back to the early history of the universe?

Above we have mentioned new interpretations (or rather usages) of genetic and structural modules; a new interpretation can also aim towards ontogeny and body structure. Some examples: loss of primary larvae; introduction of secondary larvae in some groups (e.g., insects, amphibians), or bodily transformation in parasites (e.g., the parasitic crab *Sacculina*). Also, parallel adaptations in specializations: blind cave dwellers, anoles, sticklebacks, atavisms, fish body shape (some of this in Chap. 8). Or consider speculations analogous to today's interpretation of past authors: it would be interesting to see whether the artiodactyls, in their transformation into cetaceans, drew on ancient evolutionary experience of aquatic

[5] Cloakroom tokens.

vertebrates—approached, of course, with their later, mammalian experience. However, a caveat: this theory of interpretation accepts the possibility of openness to multiple readings (in both human and non-human cultures), but it must also assume that consensus should be reached: if not on the meanings that the text supports, at least on those that it does not (see Eco, 1990).

Eco (1990 and elsewhere) again and again points out the necessity of presuppositions: we live in a small world not only in literary works, but also in ordinary communication. This means that we rely on many facets of reality that needn't be mentioned. A lot of life's situations are based on such presuppositions, and it is these that scientific inquiry is most focused on: habits established by evolution, or at least something on the level of "this is just the way it is with us". At the same time, one must assume at least a minimal *understanding* of "others" who share with me the same time and space. The world of stories is not the whole world—hence the adjective "small".

4.5 Darwin and Science

Let us stop now at the historiography (the *interpretation of the history*) of life, and let us begin, of course, with Darwin. One has read *On the Origin of Species by Means of Natural Selection*, a foundational work for modern biology, paleontology and ecology, and has reached the *last* page. There are sentences that may explain, but perhaps also obscure, all that one has learned from the voluminous book. We read: "And as natural selection works solely *by and for the good* of each being, all corporeal and mental endowments will tend to progress towards *perfection*" (Darwin, 1876: 428; emphasis ours here and below). This neatly echoes Leibniz's (2017) notion that Creation is endowed by God towards self-perfecting, self-improvement. Aristotle, who makes a similar claim (see 9.2), was apparently unknown to Darwin, but the diction is somewhat similar.

In the next and last paragraph of the *Origin of Species* we read: "These elaborately constructed forms, so different from each other, and dependent upon each other in so complex a manner, have all been *produced by laws acting around us*. Those laws, taken in the largest sense, being Growth with Reproduction; Inheritance, [...] Variability from the indirect and direct action of the conditions of life, and from use and disuse: a Ratio of Increase so high as to lead to a Struggle for Life, and as a consequence to Natural Selection, entailing Divergence of Character and the Extinction of less-improved forms" (Darwin, 1876: 429).

Both quotes refer to external *laws*, to a struggle for survival, and *then* to natural selection. The reader may, of course, object that the introduction and conclusion of any book can endure a few ornate paragraphs but let us try to take them seriously and consider the claim that the history of life is determined by natural laws equivalent to the laws of physics. It may also be argued that historicism was at its height in Darwin's time, recognizing natural laws in the history and even of the human race. But there may be a deeper reason: Darwin was a *scientist*, a Newtonian by conviction, and therefore he viewed natural phenomena through the prism of Newtonian physics. Such is also the view of authors David Depew and Bruce Weber in their remarkable book *Darwinism evolving* (1996). They write: "[Darwin] worked within a very self–conscious scientific culture in which external forces operating on bodies rather than internal drives were expected to drive change over time. Darwin now extended Newtonian model of systems dynamics to questions about biologic origins by taking a detour through economics. [...] He very nearly succeeded in extending his culture's most respected scientific model to objects and processes from which it was supposed to keep its distance" (Depew & Weber, 1996: 10). By that last sentence the authors mean that Daarwin should not became involved in *human* evolution.

And the argument and style of Darwin's book seem to suggest the opposite of what is claimed on the last page of *On the Origin of Species*. In today's language, one might rather interpret his book in a way that chance (descent with modification) and necessity (bringing offspring into the world) rule in evolution. Hence, natural selection is the "invisible hand" that keeps evolution within limits, just as such a "hand" corrects the economy or population size. However, few today would find in economics or demography a connection—other than metaphorical—with the physical forces of nature, such as gravity.

That time, which indeed craved a *scientific* explanation of evolution, could not appreciate the work much; moreover, it did not fit its spirit or worldview. We will elaborate more on this topic in Chap. 9, where we discuss, among other things, Radl's account of responses to Darwin, written 50 years after *The Origin of Species*: Radl confirms the above-mentioned interpretation of the biology of Darwin's time. As we point out elsewhere, *any* scientific explanation—Darwin's as well as his opponents—must deny the living being's innerness in the first place. And the innerness will have to be brought back into consideration, we believe, with the advent of epigenetics, such as will be specifically introduced in Chap. 10. The reader may then find an answer to the question of why there is a Sect. 4.7 on Chinese notation in this chapter.

4.6 Transformations

Loanwords are of a common occurrence, their notation is not governed by many rules, and so we have a whole host of paradoxes that are an invitation to further creation (see, e.g., Barfield, 1967). We have already mentioned, and in Chap. 5 we will discuss in more detail, what such a loan can mean—not only in the language itself, but also what happens when a word from another language is transferred to a new sociocultural environment (see also below, 4.7). The question is: to what extent such a word can capture the nature of the new socio–cultural field and to what extent can it simultaneously act constitutively and reconstructively in this new field. For example, there are often shifts in meaning, spelling, and grammar. The language of the older generation and that of contemporary adolescents of the same language area can also differ considerably, despite the painstaking teaching of "correct" spelling, grammar and speech.

The literal meaning of both utterance and notation is thus difficult to asses; moreover, it loses its literalness rather quickly—hence the endless work of hermeneutical interpretations of legal, philosophical or sacred texts; the extremes are of course coded messages, but also clichés and frozen metaphors. What is more interesting, as Barfield (e.g., 2013: 45) says, the secondary meaning of the word pertains even while the original one is still somehow unconsciously felt. However, he also points to the opposite trend, the seeking of metaphors and images, and even works of art, expressed verbatim; propositional logic is but an extreme refinement of such an opposite trend. It will be interesting to see how the sensitivity to such trends will change with the use of AI translators and stylistic tools.

Moving into the world of biology, we note that the linear notation of base order in DNA (and other biopolymers) is also a reduction to notation, hiding "body language" similar to the speech-notation doublet. We will review this information in Chaps. 8 and 9. Hence the passages on Chinese characters that follow.

4.7 Alphabet and Pictograms

Let us now compare two systems for writing speech notation: phonographic and logographic, specifically, the Latin alphabet and Chinese characters. Our so–called alphabetic script uses a linear sequence of alphabetic characters to write—usually quite approximately—the sounds of speech as they follow each other. Spoken language, however, is only partially discrete (cf. sound spectrogram, Fig. 4.1), and in addition to speech there comes also the "body language" and, of course, the contexts in which speech takes place which are inextricably linked to speech.

Letters necessarily digitize the "blurred" form of speech (and vice versa, reading the text "blurs" the writing again). Pronunciation and relative position of vowels

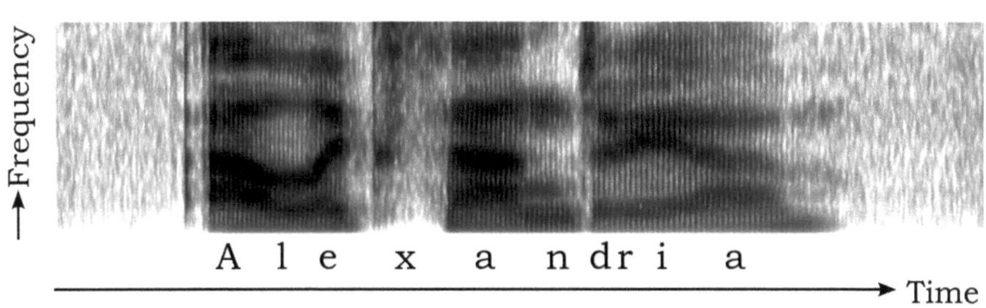

Fig. 4.1 Spectrogram and written representation of speech. The recording takes approximately 2 s, the frequency expanse about 0–3 kHz. Note the frequency resolution of the human voice and the very vague delimitation of the duration of the sounds. Let us compare the reduction of sounds to 10 letters with the much longer recording associated with the digitization of the recording (even there, of course, a reduction must be chosen that will be a compromise between the quality of the recording and the demands of the complexity of the recording). We thank D. Faltýnek for his help

also change quite rapidly—much more rapidly than the written norm, making it inappropriate and uneconomical to change the later. Trask (2010) gives examples for English: he argues that the introduction of broadcast and the fact of globalization is the only reason why the various dialects of English that are spread around the world "got used to" each other and now understand each other (while still persisting). Yet differences in spelling and vocabulary are negligible and usually do not pose a problem. We know this from our own experience: we can express ourselves in writing in a foreign language and we can read it, but most of us use a mixture of sounds and idioms that a native speaker might not be able to use. And the miracle: we will still usually be understood!

We repeat: the transcription of speech into a digital sequence of linear characters represents a significant reduction of the spoken word, made more difficult by the historical accretions of spelling and grammatical usage, but nevertheless long–lasting, copiable, storable in databases and retrievable. Here, however, comes the second part of the writing–speech relationship: reduction by writing means for the reader the necessity to interpret the digital writing back into meaning that is not digital. Random mutations of the text, but above all misunderstandings—from the most banal, spelling or grammar, to the interpretation of whole texts—together with historical shifts of vocabulary, allows for a broad fan of possible interpretations of *the same* text (not to mention translations). (Here, we will not deal with another type of speech reduction represented by recordings.)

Let us also note that in spoken and written language, Indo–Europeans primarily perceive words (and sets of words) as the smallest unit of speech; when reading we usually recognize whole words, that's why we so often miss typos, etc. For the sake of further explanation let us, however, recall the existence of word *roots* and their prefixes, suffixes, infixes, etc.; we must learn these for various entrance exams; after we pass, their knowledge does not matter so much. Here, however, it must be stressed—especially in connection with the discussion below—that there has never existed any original, primitive language composed only of roots that our ancestors barked at each other: "'Roots', far from being the germs of speech, are the product of ages of intellectual abstraction carried on, first, instinctively by ordinary speakers, and afterwards deliberately by the grammarians and philologists. The service rendered by these latter both to speech and to thought is of immense importance; their error merely lay in supposing that life actually created language after the manner in which their logic reconstructed it. They mistook elements for seeds—and called them roots." (Barfield, 1973: 81–82).

And one more lesson from Barfield's work (1973): on the one hand language is a historical creation of the community (i.e., Closure 2), its forms have always *somehow* reflected the *collective participation of the community* in the world, a way to understand the world (how to construct an idea of it), and on the other hand, the world somehow imposes and imprints itself on the collective consciousness, through the filter that language represents.

Let us now compare this with the speech–writing relationship of Chinese (and some East Asian peoples; for simplicity, let us call it "Chinese".[6] Here, the previously mentioned word roots, mostly monosyllabic, play a much larger role; moreover, these syllables can be pronounced—in standard mandarin—with four kinds of intonation, thus denoting quite different things. The notation usually consists of a radical, which establishes a certain range of meaning determination, and a phonetic component, which indicates pronunciation; yet the intonation of pronunciation cannot be guessed from the character itself. For the sake of conceptualization, Fig. 4.2 (left) shows the possibilities of the two-syllable notation, its pronunciation and translation into English.

The number of radicals is a little over 200 and the number of phonetic characters is perhaps 80,000: for ordinary conversation and written communication, 2000 will suffice, but an educated person can handle 5000–8000. To make things less simple, not every character is accompanied by a radical, not all possible syllables are used in Chinese, and not every syllable is pronounced in all four ways. In addition, phonetics may have shifted over time, and pronunciation is now based on convention for most characters. This entire system was established in the centuries before BCE, although the specific character designs (notation) and pronunciation conventions have changed over time.

It is said that pictograms are universally understood by all nations that use this kind of writing. This is only partially true, and for very old characters only; it is not true at all for pronunciation, see below. Moreover, many Chinese dialects have no script at all, and children must learn both standard Chinese and the notation and pronunciation of its characters at school. By the 1950s, the situation was becoming untenable, and there were efforts to switch to the alphabet (*pinyin* script); the advent

忧郁	[yōu yù]	melancholy
犹豫	[yóu yù]	hesitation
由于	[yóu yú]	due to
鱿鱼	[yóu yú]	squid
有余	[yŏu yú]	in excess of
优于	[yōu yú]	better than

Fig. 4.2 On the right, an illustration of the relationship between the notation and sound of a two-syllable word (taken from the Internet). On the left, a notation of the scientific term "entropy" (after Mackay, 2001). Discussed in the text

[6] We would like to thank K. Gajdošová for introducing us to this issue; needless to say, if our depiction in the text is somehow distorted, it is to our credit. We have drawn also from a short commentary by Mackay (2001), and textbooks for beginners.

of computers, however, has halted this development. And yet literate Chinese are familiar with this alphabetic script: they tap it into their mobile phones which then offers them one or more convenient characters to choose, depending on the context.

Let's follow the examples of loanwords (examples from K. Gajdošová's letter): "When creating new words, Chinese behaves similarly to European languages—it either adopts the phonetic form of a foreign word, or it goes after the meaning, or it mixes it in one word, or both coexist before one of them takes over. Thus, the Internet is *hu lian wang* [network-of-inter-connections], vitamin is *wei sheng su,* [maintain-life-ingredient], phenomenology is *xian xiang xue* [science-of-appearance-of-images]. A nice example is evolution: *jin hua* [going-forward-changing]. At the same time, radar is *lei da* (depending on how Chinese manages to mimic English pronunciation); engine is either *ji dong* [movement-machine] or *mo tuo* or *yin qing*; beer is *pi jiu*, which is the Chinese reading of the Japanese characters *bi-ru* (i.e., beer), etc. The character for "philosophy", read in Chinese as *zhe xue*, is just a literal reading of the Japanese characters *tetsu-gaku* [wisdom-science]. Then, for example, a vector might be *xiang liang* [direction-size/measure], or *shi liang* [arrow-size], or *fen liang* [component-size], etc."

It can be added, for example, that local names are written in characters whose pronunciation even Europeans familiar with Chinese and Chinese writing may have no idea of. At the train station, even if they know the name of their destination, they are lost at the departure board and must ask the locals for the pronunciation. (Or vice versa: they hear the announcements and have no idea where the train is going until they see the notation.) Understanding notation is thus apparently much looser in Chinese than in alphabetic notation: Chinese encompasses a larger context of notation, and in reading (and speaking) it is said to adhere more closely to established clichés that prevent an overly large proliferation of interpretations.

What follows is an excursion into scientific terminology, as Mackay (2001) brings up with the word *entropy* (Fig. 4.2, right). "Thus, the character in the figure has the fire radical on the left; on the right it has, as phonetic, the character for 'merchant', which is pronounced *shang*. The character as a whole means entropy, and is also pronounced *shang*. [...] In spite of appearing to be a stream of single characters, the Chinese language consists mainly of polysyllabic words made up of several characters.[7] With a modern word processor, if you type in a single syllable, the program will offer you perhaps a dozen characters all pronounced in the same way. You could choose one of them, but if you continue to type successive phonetic syllables of the word or words in a phrase, there is usually a unique output of characters. Entering the phonetic *shang* brings up eleven characters, with 'entropy' as the least likely choice. The enunciation of the phonetic *shang* by itself is thus not enough to bring up the meaning 'entropy', so a teacher has to write the character on

[7] The question is whether we should not also take inspiration from this form of notation in biology. We think of genetic notation as the linear writing of "letters" in DNA, but DNA is embodied in *chromatin* of a specific and variable shape that largely determines the context of gene expression. See Glossary and Chap. 10.

the board or, in conversation, launch some explanation of what is being talked about. Nevertheless, at a Chinese version of C. P. Snow's cocktail parties, people can talk about whatever may be necessary, exploring each other's knowledge or ignorance, and later repair any deficiencies using the Chinese Internet. People remember characters according to their real or fictional etymology, so linking fire or energy to buying and selling is a convenient mnemonic device if you want to write 'entropy'".

Despite such differences, the lessons to be learned from comparing the two spelling systems are quite similar. *All* languages can make do without reduction to writing, though then communication is heavily dependent on memory and fixed phrases (the meaning of which can, however, change with time: how do we *know* that *today's* form of mythology reflects the thinking of the ancients?) Such reduction, however, opens up new worlds which, when read, then fit into the reader's understanding of the world; hence, in fact, this text of ours.

Now, a comparison of both writing systems follows. Despite the different ways of writing and due to the conservative spelling of alphabet-using nations, the two systems seem to converge: pinyin on the one hand, and "pictograms" on the other. Concerning English, the contemporary *lingua franca* of the world, Trask (2010: 152) states:

> Pronunciation, like every aspect of English, continues to change [...] So, even if we agreed on the spelling that worked well this year, that same spelling might after a while be as ridiculous-looking as is our spelling of *knight*—which at one time made perfect sense. It appears, therefore, that we are drifting slowly towards a writing system rather like the one the Chinese use. The Chinese script, there is no attempt to write the pronunciation at all. Instead, each individual word is conventionally represented, by an arbitrary set of symbols, and learning the particular set of symbols associated with every word in the language [...] In our case, we simply have to learn that squiggles like *reign* and *missile* and *doubt* stand for particular words with particular meanings [...] Those speakers in England for whom *tyre* and *tower* are both pronunciated like *tar* (and also *ta*, [thanks]) will just have to master the squiggles required for each of the four meanings, much as Chinese speakers have to learn the different squiggles that represent words of identical pronunciation but different meanings.

Let's add that we can also introduce diacritics, which better suggest the correct pronunciation—as most alphabet-writing nations do, albeit to varying degrees. We shall also encounter diacritics, and we think for the same reasons, at different levels of the living (see Chaps. 8 and 9).

The shape of the *present* globalized world (its science, technology, and ways of living in general) is still largely the result of the efforts of nations that have adopted alphabetic writing: it allows for easier development of notation, calculus, logic and science. The linear sequence of discrete characters (as it has been since antiquity) has also grown into our vision of the world. Perhaps this is why the structure of the world as a composite of a finite number of basic elements is somehow perceived in this way by physics, chemistry and biology. However, this may not be a permanent form: it remains valid only until nations based on "Chinese" notation discontinue adopting the mechanism of our achievement and begin again making their own unique contribution to the common knowledge based on their own heritage of notation.

And why this whole chapter in the first place? In the twentieth century, evolutionary theory was largely focused on notation—strings of alphabetic characters. From these genetic notations, biology—in combination with molecular biology—then attempted to explain both the nature of life and its evolution (indeed, its model was an only slightly older one from comparative linguistic). In short, the story goes, genetic notation perhaps determines, even unambiguously, what kind of cell or multicellular body will arise. Chapters 7, 8, 9 and 10 will show what this reduction to notation has meant for the understanding of the living in biology. Explaining the forms of living bodies directly by machine translation of genetic *notation* represented the ethos of life interpretation of the past several decades; changes were attributed only to *accidental* changes in this notation (e.g., Monod, 1972). The advent of epigenetics (Chap. 8) changes this situation; we discuss the role of molecular "accentuation" as an analogue of epigenetic changes there.

References

Assmann, J. (2011). *[1997] Cultural Memory and Early Civilization: Writing, Remembrance, and Political Imagination*. Cambridge University Press.
Barfield, O. (1967). *[1926] History in English words*. Lindisfarne Books.
Barfield, O. (1973). *[1928] Poetic diction. A study of meaning*. Wesleyan University Press.
Barfield, O. (1988). *[1957] Saving the appearances. A study in idolatry*. Wesleyan University Press.
Barfield, O. (2013). *[1977] The rediscovery of meaning and other essays*. Barfield Press.
Calvino, I. (1981). *If on a winter's night a traveller*. Harcourt Brace Jovanovich.
Carrière, J. C., & Eco, U. (2012). *[2009] This is not the end of the book*. Northwestern university Press.
Carroll, S. B. (2006). *Endless forms most beautiful: The new science of evo devo*. Norton & Co..
Carroll, S. B., Grenier, J. K., & Weatherbee, S. D. (2001). *From DNA to diversity: Molecular genetics and the evolution of animal design*. Blackwell.
Cowley, S. J. (Ed.). (2011). *Distributed Language*. John Benjamins.
Cowley, S. J., & Markoš, A. (2019). Evolution, lineages and human language. *Language Sciences, 71*, 8–18.
Cowley, S. J., & Valée-Tourangeau, F. (Eds.). (2013). *Cognition beyond the brain. Computation, interactivity and human artifice*. Springer.
Darwin, C. (1876). *[1857] Origin of species by means of natural selection* (6th ed.). John Murray.
Depew, D. J., & Weber, B. H. (1996). *Darwinism evolving. Systems dynamics and the genealogy of natural selection*. MIT Press.
Eco, U. (1989). *[1967] The open work*. Harvard University Press.
Eco, U. (1990). *The limits of interpretation*. Indiana University Press.
Eco, U. (1994). *Six Walks in the Fictional Woods*. Harvard University Press.
Eco, U. (1997). *The search for the perfect language*. Blackwell.
Eco, U. (2014). *[2007] From the tree to the labyrinth historical studies on the sign and interpretation*. Harvard University Press.
Flegr, J. (2010). Elastic, not plastic species: Frozen plasticity theory and the origin of adaptive evolution in sexually reproducing organisms. *Biology Direct, 5*, 2.
Fromm, E. (2013). *[1966] You Shall Be as Gods: A Radical Interpretation of the Old Testament and Its Tradition*. Open Road Media.
Gilbert, S. F. (1994). *Developmental biology*. Sinauer.

Ginzburg, C. (1989). [1980] Clues: Roots of an evidential paradigm. In *Clues, myths, and the historical method* (pp. 87–113). Johns Hopkins University Press.

Ginzburg, C. (2013). *The Night Battles: Witchcraft and Agrarian Cults in the Sixteenth and Seventeenth Centuries*. Johns Hopkins University Press.

Gurevich, A. (1990). *Medieval popular culture*. Cambridge University Press.

Halbwachs, M. (1980). *[1950] The collective memory*. Harper & Row.

Harris, R. (2009). *Rationality and the literate mind*. Routlege.

Koselleck, R. (2004). *[1979] Futures past: on the semantics of historical time*. Columbia University Press.

Kozmik, Z., et al. (2008). Assembly of thc cnidarian camera-type eye from vertebrate-like components. *PNAS, 105*, 8989–8993.

Leibniz, G. W. (2017). *[1710] Theodicy: Essays on the Goodness of God, the Freedom of Man, and the Origin of Evil*. Create Space Independent Publishing Platform.

Lotman, Y. M. (2009). *[1992] Culture and explosion*. Mouton de Gruyter.

Love, N. (2004). Cognition and the language myth. *Language Science, 26*, 525–544.

Love, N. (2007). Are languages digital codes? *Language Science, 29*, 690–709.

Mackay, A. L. (2001). Character–building. How do writing systems such as China's deal with the 21th century? *Nature, 410*, 19.

Markoš, A., & Švorcová, J. (2019). *Epigenetic processes and the evolution of life*. CRC Press.

Markoš, A., et al. (2013). Living as languaging: distributed knowledge in living beings. In S. J. Cowley & F. Valée-Tourangeau (Eds.), *Cognition beyond the brain. Computation, interactivity and human artifice* (pp. 71–92). Springer.

Merton, R. L. (1967). On the history and systematics of sociological theory. In *On theoretical sociology. Five essays, old and new*. Simon & Schuster.

Monod, J. (1972). *[1970] Chance and necessity. An essay on the natural philosophy of modern biology*. W. Collins Sons & Co.

O'Connor, M., Garm, A., & Nilsson, D. E. (2009). Structure and optics of the eyes of the box jellyfish *Chiropsella bronzie*. *Journal of Comparative Physiology A, 195*, 557–569.

Rappaport, R. A. (2010). *[1999].Ritual and religion in the making of humanity*. Cambridge, Cambridge University Press.

Švorcová, J., et al. (2018). Origins of cellular biosphere. In V. P. Sahi & F. Baluška (Eds.), *Concepts in cell biology – history and evolution* (pp. 271–290). Springer Int.

Trask, R. L. (2010). *Why do languages change?* Cambridge University Press.

Vansina, J. (1985). *Oral tradition as history*. University Wisconsin Press.

Zweig, S. (1942). *The world of yesterday*. Viking Press.

Chapter 5
The Disenchantment of the World

Abstract The European Reformation of 1520s disenchanted the Sacraments; from the point of semiotics, what occurred was separation of the *representamen* from the *object*. For Weber, it meant one of the "big events" in the process of disenchantment of the world. We compare the disenchantment theory with the theory of "illumination of the spirit" (Cassirer), or with overcoming the "original participation" (Barfield). The evaluation of such an event was formulated by authorized scholars as a minimum of "before" and "after" constituting the unity that distinguishes an event from mere incidents (Koselleck). Such "before" and „after" depend on the *construction* of possible history by the particular scholar. Hence, for Weber the "after"—of the disenchantment of the Sacrament and "demagicizing" the world in Reformation—indicates a plastic transformation of the culture, whereas in Barfield it may be taken as a mere elastic incident, a divergence of cultural development.

Keywords Big event · Sacraments and Lord's supper · "Before" and "After" · Plastic and elastic phases of (cultural) evolution · Magic and religion · Horizon of expectation

> We can give any meaning to events, suppressing some and picking up others, by making one circumstance be the cause of another. There is no path ahead of us except where we have just poured it for the next step we are about to take, trusting firmly that we will not crash into the void. So events have happened, but we don't really know how; we don't even know if the things we notice most were the important things in them, or if they were important at all. [...] But once the story is finished, once it is constituted in our narrative, it takes on a meaning that retroactively legitimizes all the arbitrary steps we took in creating it. [...] The important thing is to keep asking what if... We don't have to include these ifs explicitly in the model (and we probably won't), but they have to be present in it, otherwise it loses its meaning. (Dušan Třeštík, 2003, p. 34)

A. Markoš, J. Horský, *The Seeds of Forms and the Forms of Seeds*, Biosemiotics 12, https://doi.org/10.1007/978-3-032-16131-4_5

The introduction of the previous chapter works with the concept of "explosion" (upsurge), during which, according to Yuri Lotman, the dominant elements within the development of the culture system change and the curve of cultural development jumps to a completely new trajectory. The signs of such an "explosion" are shown by the changes that first took place in the German–speaking world in the 1520s, which Peter Burke (1990) interprets using the "theory of incongruence". Reformation theologians and preachers of the time revealed the existing sacramental symbolism as empty, it was—in cultural anthropological terms—"demagicized". In the magical worldview, representamen (R) and object (O) merge into one (for terminology, see 3.2 and Fig. 3.1.): the ceremony at the altar with the sacramental bread and wine, the Catholic Church states that the Eucharist is the body and blood of Christ under the species of bread and wine. In the magically experienced world, the one who manipulates the component R (bread and wine) in conjunction with the component O (flesh and blood of the Lord), can by the grace of God make it immediately corporeal, graspable with the mouth. To assert that this is not so, that component R only symbolically represents component O but is ontically separated from it, can cause an "explosion". The Reformation was thus (e.g., for Richard van Dülmen, 1987) also a social revolution (just as in the previous century it was Hussitism; Šmahel, 2002, 2006). On the one hand, when we look at the beginnings of the European Reformation, such a demagicizing, or even disenchanting, event seems obvious. And yet on the other hand, we should immediately point out that we cannot in any way claim to give an immediate account of the "events" that had taken place in the minds of the actors.

5.1 Finding Unity in Diversity

Frank R. Ankersmit praises both historicism and contemporary narrativism, since both recognize "that the historian's task is essentially interpretative (i.e. to find unity in diversity)" (Ankersmit, 1994, p. 35; see Chap. 2). However, these two strands differ in the ontological validity they claim for their interpretations. "Historians attempted to discover the essence, or, as they called it, the *historische Idee*, which they assumed was present in the historical phenomena themselves. Narrativism, on the contrary, recognized that a historical interpretation projects a structure onto the past and does not discover it as if this structure existed in the past itself" (Ankersmit, 1994, p. 36). Moreover, narrativist critiques of historiography, since Ankersmit wrote these words, have gradually progressed from addressing the issue of the (metaphorical) representation of the "past" in historical narrative interpretations (e.g., Ankersmit, 1994) to the non–representationalism advocated by some authors today (Zeleňák, 2015; Kuukkanen, 2015).

The "cultural sciences" according to Max Weber, one of the classics of the noetic in historical sciences, do not reveal the "factual" contexts of "objects", but only the mental contexts of problems. According to Weber, these contexts are ascertained by means of ideal–typical modelling. "In its conceptual purity, this mental image cannot be found empirically anywhere in reality, it is a utopia, and for historical work the task arises of determining in each individual case how close or how far reality is from that ideal image" (Weber, 1904, p. 65).

Weber's reliance on the possibility of determining the difference of selected parts of reality with respect to an ideal–typically constructed or modelled image is epistemologically more optimistic than Ankersmit's position quoted above. Despite this, however, it is true that both the logic of interpretation by such construction (Weber 1904) and the logic of narration are strictly nominalist (Ankersmit, 1994, p. 39).

As opposed to mere historiography (*Geschichtsschreibung*), the science of history (*Geschichtswissenschaft*) seeks to establish a meaningful unity in diversity by introducing a system of *diachronic and synchronic connections* or *relations* into a particular set of source traces ("facts", research–worthy "problems") through its theories or narrative interpretations. The essential question is whether the relations thus interpretatively established have their ontic counterpart in the form of a real mutual configuration of the individual elements of the socio–cultural reality under study. If we start from the idea of a fundamental separation of the sphere of "culture" from the sphere of "life", it then seems that the developmental lines and relational networks between observed phenomena interpretatively "found" by historians of culture are established much more (if not completely) arbitrarily than those established by the analogous procedures by evolutionary biologists. We, however, choose a different perspective. As we see it, the (supposed) gap between "life" and "culture" is bridged by categories such as "experience", "memory", "sign", or "Closure 2".

For, as we show in this book, "culture" cannot be separated sharply from "nature", nor can it be set in stark opposition to the the biosphere; this because "culture" grows organically out of biological evolution. For example, Jacques Monod (1972, p. 128) considers "the development of human symbolic language" as "a unique event in the biosphere" that opened "the way for a second evolution, the creator of a new realm: the realm of culture, ideas and knowledge".

Seen in this way, it is at least possible to conceive of culture as one of the many tools of human adaptivity. Admittedly, the mechanism is (semi)autonomous, and as such it can even become maladaptive (Richerson & Boyd, 2006, pp. 218–279). However, and precisely as a consequence of the [semi-]autonomy of "culture", we must add that what is maladaptive in terms of biological evolution need not be maladaptive in terms of cultural evolution. A stronger argument against the sharpening of the divide between "culture" and "biosphere" is then provided by biosemiotics, or semiotics in general, including the semiotic concept of culture (Cassirer, 1946, 1955, 2020; Geertz, 1977). Within this framework, it can even be said that "culture" —understood as a semiotic system—has existed since the very beginning of life, and therefore it is of no surprise that from the very beginning culture has coevolved with genes. Returning to the historical sciences, from this perspective of ours the differences between cultural history and evolutionary biology may appear not as differences in kind, but only in degree.

5.2 "Event" Theory: "Big History" and "Big Events"

In recent years, there has been a growing body of work that claims to be "big history" (Krakauer et al., 2017) that seeks to integrate the history of humanity into cosmic or planetary history. In doing so, it discusses developmental lineages for which—while not in principle resisting a constructivist approach—claims to some extent some sense of greater ontological validity than Weber or Ankersmit would have done. A "big history" conceived in this way could be characterized by its *integrity* (or complexity; Spier, 2017) and, at the same time, by its capacity to identify turning points or watershed events. If we consider "events" within the historical process, it is appropriate to ask whether this process is meant to be an *all–encompassing historical process* (in the sense of Hegel's objectification of Spirit) or a partial process of the transformation of a defined *entity*. Even if we mean it as an all–encompassing historical process, it would be necessary to clarify whether we are referring to "big history" in the sense of (a) the history of humans (both in the evolutionary–anthropological and socio–cultural sense) or in the sense of (b) the history of civilizations, societies, and cultures ("quasi–characters"; Ricoeur, 1984, pp. 197–205). Finally, however, it could also be (c) the history of "great", albeit quite unique, individual political, religious, economic "events" that defined an epoch or changed the course of previous developments.

The course of an entire history, possibly conceived as an all–encompassing historical process, can only be described historically by a narrative. This is all the truer, the greater the ambition for complexity of big history (Spier, 2017). However, we have no theory of the whole of history, let alone any (logically coherent) law of history that can be formulated. It is, however, possible to think of formulating theories of partial historical processes, theories of the transformation of certain entities (Rüsen, 2006).

To be historical is to be in time and in change, spanning between the future and the past (Angehrn, 2004, p. 392). "Change", however, requires an "event"; in other words, an "event" by definition induces change. From the perspective of the constructivist approach, however, events are not the ontic building blocks of history, they are constituted only in mental research operations. "A minimum of 'before' and 'after' constitutes the significant unity that makes an event out of incidents" (Koselleck, 2004, p. 106). What Koselleck refers to as a "collection of incidents" (*Begebenheiten*) can also be thought of as a "configuration of traces" (Ginzburg, 1989). The possibility of establishing a minimum of "before" and "after", however, requires having at least an implicit idea of some "history", or a certain "sequence" of them, which is a priori with respect to the relevant "event". Koselleck speaks of "a theory of possible history": "Such a theory is implicit in all the works of historiography; it is only a matter of making it explicit. There is a wide variety of statements on history in its entirety or individual histories which cannot be directly related to the sources, at least in the second phase of study" (Koselleck, 2004, p. 150). He goes on to say: "In principle, a source can never tell us what we ought to say. It does prevent us form making statements that we should not make. The sources have the power of veto." (Koselleck, 2004, p. 151).

However, in relation to one and the same "collection of incidents" we can very likely formulate multiple historical sequences (sections of developmental lineages), which will always establish different minima of "before" and "after". One and the same "collection of incidents" can thus be considered as several different evolutionary or historical "events". If these "incidents" are parts of "theories of possible histories", all of which can withstand a "veto of the sources", then it will have to be admitted that they are all equally valid. We thus get parallel pictures of "(great) histories" with different (breakthrough) "events" contained within them.

Koselleck speaks of a "theory" of possible histories. However, these are rather hypothetical interpretative ideas that presuppose or involve an (implicit) conception of "history". The "theory of possible histories", as Koselleck understands it, only sometimes has the features of a historical–scientific "theory" in the narrower sense (e.g., Wehler, 1979; Welskopp, 2008). In other cases, it involves sequences or plots from broader historical narrative interpretations (Mink, 1970; Ricoeur, 1984; Ankersmit, 1994).

The parallelism of "theories of possible history" and the variety of "events" constituted within them raises the question of whether "theories of possible history" are culturally derived or value–neutral. Certainly, they are culturally colored in the sense of the noetic operation of "value relation" (*Wertbeziehung*) analyzed by Weber. However, they should not—in Weber's conception of the "value neutrality" of science—be evaluated as "desirable or undesirable" (Weber, 1991, pp. 176–236; Müller & Sigmund, 2014, p. 150). In spite of this, even in Weber's intentions, we could not exclude a "value judgment" (*Werturteil*) when discussing a certain "collection of incidents" as an "event" (Weber, 1991, pp. 111–115), i.e., a certain researcher's sense of value belonging to the actors.

If we extend the noetic perspective of historical sciences to include approaches from the sociology of knowledge or the socio–cultural anthropology of memory, we can also think of major historical events as "symbolic centers". Miloš Havelka (1999) links Rickert's notion of the "historical center" with Cassirer's theory of "symbolic formation" and offers a theory of "symbolic centers". According to the latter, relativized "historical centers" can take on many theoretical and general cultural, rational and emotional symbolic meanings. "Symbolic formation" introduces a more general meaning, which, however, can take very diverse, even contradictory forms in the same socio–cultural field. They thus become nodal points in the construction of our image of history as well as the construction of collective memory. We can cite examples of different types of "great historical events" and, where appropriate, of how they can be perceived as "symbolic centers".

All of these "big", "historical" "events" can be discussed within the framework of admittedly culturally based but value–neutral "theories of possible history". Sometimes, however, we find that these "events" represent intersections of competing "theories of possible history" or competing "historical narrative interpretations". For example, the radical fertility decline of the past few decades may appear maladaptive from the perspective of evolutionary anthropology, as it reduces human biological fitness (as measured by reproductive success; Richerson & Boyd, 2006). However, when viewed from the perspective of cultural evolution, even this significant drop in fertility may appear adaptive. Indeed, we can contrast the biological criterion of maladaptation with the criterion of cultural maladaptation. Maladaptive, from the point of view of cultural evolution, is that which leads to a collapse of cultural transmission and a decrease in cultural diversity. From this perspective, it is possible that contemporary culture requires much more time for its transmission by social learning, including the time of educated women; and thus, one might say in opposition to Richerson and Boyd, it requires greatly reduced fertility. In terms of the ecological crises of the Anthropocene (Tamm & Simon, 2020), the limitation of population growth may appear as adaptive behavior of the human species.

But these "events" may also come to be—at different times (Tamm & Simon, 2020) differently viewed—"symbolic centers". However, as "symbolic centers" thet already lose their "value neutrality". For example, the "industrial revolution" may appear as a "heroic enterprise" from the perspective of the mid–twentieth century, but from the perspective of the current climate crisis and efforts to limit CO_2 emissions it may seem more like the beginning of the Anthropocene crisis. Similarly, the demographic transition (fertility decline) will appear as a differently connoted "event" within competing theories and narratives. As a "symbolic center" it will thus lose its value–neutrality. When we try to integrate above–mentioned "events" (e.g., the "de–magicisation of the sacraments", the "emergence of modern natural science", the "industrial revolution", the "demographic transformation") into the "great history" of humans or human civilization(s), we do so much more as value–connoted "symbolic centers" than as value–neutral "events" (support for this claim: Ankersmit, 1994; Ricoeur, 1984).

In the context of what has just been said, we will examine breakthrough "events" within the process of "disenchantment of the world" in more detail. This will open the possibility of distinguishing plastic and elastic phases within the framework of cultural development (i.e., to the question of analogy between the practices of cultural historians and evolutionary biologists).

5.3 "Illumination of the Spirit" and Overcoming of "Original Participation"

According to Max Weber, the process of the disenchantment of the world is manifested in the Christian religious sphere, among other things, by the gradual rejection of magical means of salvation. Weber sees the turning points in the teachings of the

Old Testament prophets, in Hellenistic scholarship, in the critical attitude of the sixteenth–century European Reformation toward the Catholic concept of salvation, and especially in the symbolic conception of the sacraments in Protestantism (Weber, 2005; Havelka, 2012). If we then connect Weber's reflections with Cassirer's interpretation, it turns out that the disenchantment of the world could (perhaps) be explained as one instance of a more general developmental scheme (or evolutionary principle). Ernst Cassirer considers the "gradual illumination of the spirit"[1] to be the process and nature of the gradual differentiation of the contents of spirit (individual symbols and whole symbolic forms) in which new spiritual "dimensions" emerge. At the same time, it takes form in the emergence of different ways of dealing with these contents of the spirit. Beginning with the "primordial division" into the sacred and the profane (Cassirer, 1955, p. 70), new spiritual entities are constituted by further, increasingly subtle differentiations. At the same time, the ways of dealing with these entities also change. "The new ideality, the new spiritual dimension, that is opened up through religion not only lends myth a new signification but actually introduces the opposition between 'meaning' and 'existence' into the realm of myth" (Cassirer, 1955, p. 239). The separation of "existence" and "meaning", i.e., the separation of "object" and "representamen" is in fact a de–magicisation of symbols, a realization that they are mere signs (meanings) denoting essentially separate existence.

Owen Barfield also discusses a similar change when he contrasts the "texture of medieval thought" and "modern thought". The transition from medieval to modern thought, in Barfield's terms, is related to the overcoming of "original participation" (see also below). "For non–participating consciousness, it is either–or. A narrative is *either* a historical record, *or* a symbolical representation. It cannot be both" (Barfield, 1988, p. 149). The Middle Ages, by contrast, do not bring literal and symbolic meaning into strict opposition (Barfield, 1988, pp. 74 and 84–91). Barfield's interpretation corresponds, for example, with Huizinga's (1996) thesis on medieval symbolism and the transition from "symbolic thinking" to "causal thinking" or with Burke's use of Huizinga thesis by linking it with Geertz's theory of incongruence.

According to Burke (1990), incongruence is manifested through the emptying of traditional symbolism. Burke finds a break in this emptying in the Reformation critique of the sacraments in the 1520s. The difference between medieval and modern thought is then also described by Barfield using the example of the Eucharist. He says:

> For all who partake of the Eucharist first acknowledge that the man who was born in Bethlehem was 'of one substance with the Father by whom all things were made'; and then

[1] We attempt a synthesis with the evolution of life, and invite the reader to understand "spirit" with the more neutral "innerness" as we defined it in the Introduction, a property determined by two types of Closures. Then the "immanent development of spirit" in human history becomes analogous to the "gradual increase of complexity" in biological evolution, as many evolutionists operate with this idea. Note (in both cases) the absence of criteria for evaluating this development — if only because it is not easy to define spirit or complexity, let alone the degree of their development or differentiation.

> they take that substance into themselves, together with its representations named bread and wine. That is after all the heart of the matter. There was no difficulty in understanding it, as long as enough of the old participating consciousness survived. It was only as this faded, it was only as a 'substance' behind the appearances gradually ceased to be an experience and dimmed to a hypothesis or a credo, that the difficulties and doctrinal disputes concerning trans–substantiation began to grow (Barfield, 1988, p. 170).

5.4 The Lord's Supper in the Reformation Faith

In High Medieval scholastic doctrine, the dogma of the transubstantiation of the sacramental bread and wine into the flesh and blood of the Lord distinguished a twofold level of communion. It was received both "spiritually" (*spiritualiter*) and "sacramentally" (*sacramentaliter*). Spiritually we receive by faith, by the action of the Holy Spirit. Everyone who is present at the celebration of the Lord's Supper and has faith receives spiritually in the moment the priest pronounces the consecration formula.

"Sacramentally", the bread and wine ("elements of the sacrament") are received with the bodily mouth. In sacramental communion, the priest may stand in for the laity. The original doctrine was that it is the spiritual communion that is the saving communion. As the magical worldview was strengthened during the fourteenth century, some Reformed preachers (e.g., Matthew of Janov) began to emphasize the "sacramental" communion (even considering it the 'source' of salvation) and stressed that it should be accessible to lay people more often. In connection with the emphasis on the saving character of sacramental communion, the Hussite demand for a lay chalice for communion under both kinds, even for children, is understandable (Kybal, 1905, pp. 202–227 and 298–318). At the same time, there were disputes as to whether the "elements of the sacrament" (i.e., the body and blood of the Lord) are "substantially" present in the bread and the wine, or whether even after the priestly act of transubstantiation the bread remains the bread and the wine the wine (doctrine of remanence; Kybal, 1905; Hanák, 1928–1929, pp. 1–44; Šmahel, 2002, 2006).

The notion of the saving character of sacramental communion is actually a notion of the sensual or bodily grasp of the grace of God and its material manipulability. From the perspective of at least some medieval actors, the grace of God and its saving power is—in the terminology of the semiotic theory of the sign—directly in the vehicle or in the representamen. It is this notion that the sixteenth century Reformers had in mind when they criticized the magical conception of the sacrament in the Catholic church.

Within the framework of these criticisms, for example, the Unity of the Brethren developed a theological doctrine which, on the one hand, rejected the idea that Christ (or rather his body and blood) is present in the Lord's Supper "substantially", i.e. "bodily", "earthly", but on the other hand they reject also that he is present in it only "symbolically", i.e. "signified" or "figuratively". They chose a kind of "middle way" and taught that Christ was in the Lord's Supper "spiritually" and "sacredly", that is, "truly" and "powerfully". "Spiritually" we receive Him by faith, "sacredly" by the mouth. This doctrine, first systematically formulated in 1526, then permeates the Unity of the Brethren texts until the twentieth century (Odložilík, 1923, p. 124). Insofar as the Reformers emphasized "spiritual" communion, they actually returned, in a sense, to what the Catholic Church had taught before the above mentioned "magicizing" conception of the sacrament, namely, that saving power has the communion on the "spiritual plane", not on the "sacramental" plane.

The doctrine of transubstantiation is one of the products of the scholastic Thomistic rationality of the thirteenth century. Since the 1620s—thanks to Luther and especially to Zwingli—it has become one of the main points of the Reformation critique. Zwingli, in particular, was very much troubled by the idea that "essence" could be directly received ("with the mouth of the flesh"), experienced by the bodily ingestion of those "representations—bread (the Host) and wine", i.e., the transubstantiation of the transfigured body and blood of the Lord. Against the magical conception of the saving power of the Eucharist, Protestantism offered either a weakened or a symbolic conception of the sacraments. The formulations of the Philippist Confession of 1566, adopted at a gathering of the Neo–Utraquistic clergy in Prostějov (Moravia), may serve as an example.

The drafting and publication of this confession is related to the way in which the Moravian nobility began to deal with the religious situation on their estates in the second half of the sixteenth century. At that time there was no legal guarantee of religious freedom in Moravia. However, the Moravian Provincial Assembly repeatedly resolved (from the 1520s onwards) that faith was a gift of God and that no one could be forced to it (in specific confessional forms) *ex officio*. The evangelical superiors on their estates began to act similarly to the German princes, on the level of individual counties—i.e., they explicitly or implicitly adopted the model of the so–called *Landeskirche* (Land Church). In this system, the landowner (in Moravia, the owner of the estate) took the lead of the local church and began to regulate the religious conditions in his territory. Among other things, the superiors began to issue normative religious texts for their subjects or give space to the clergy who were in the parishes of their estates to write such texts. One interesting example is the *Philippist Confession*.

The Philippist Confession says of the sacraments that they are "visible signs of the invisible grace of the Divine". The Lord's Supper, then "is the sacrament of the true body and true blood of Christ, depending on two things: The first, which is seen, the second, which is not seen, but heard of through the word of God, instituted and ordained for the remembrance of the sufferings and death of Christ undergone for our sins, for the confirmation of faith in us, for the spiritual saturation of our souls, and for the distinction of the society by which we are united to Christ the Lord and his holy Church" (Hrejsa, 1912, p. 36). This denomination is "Philippist" (as Hrejsa calls it) because it is religiously consonant with "Philippism", one of the two main streams into which Lutheranism split during the sixteenth century in the German and Central European region. Lutheranism split into the Flacian and the Philippist movements, among other things, precisely as a result of the controversy over the sacrament of the Lord's Supper. The text of the basic Lutheran confession of faith, the Augsburg Confession (the so–called Augustana), was written in 1530 by Philipp Melanchthon. He included only a very brief article on the Lord's Supper, "Of the Supper of the Lord they teach that the Body and Blood of Christ are truly present, and are distributed to those who eat the Supper of the Lord; and they reject those that teach otherwise" (https://bookofconcord.org/augsburg–confession/#ac–x–0001).

The disputed word was not "truly", but (not included in the quotation above) "substantially": the issue was whether people could take the body and blood of the Lord "substantially" in the Lord's Supper. Martin Luther insisted that they could. However, contrary to the existing Catholic understanding, he judged that this was not due to the transubstantiation act done by a priest, for to admit this would leave priests and the ecclesiastical institution free to "treat Christ according to their will". This happens, according to Luther, intrinsically by the power of (individual) faith *in the heart* of each person. Such a conception, however, had been rejected since the 1520s, especially by Huldrych Zwingli, by the so–called Upper German or Swiss Reformation. Zwingli came up with a radically symbolic conception of the sacraments (Blickle, 1992). Melanchthon was closer to this symbolic conception of the sacraments: as can be seen above, Melanchthon did not use the word "substantially" in the text of the *Augsburg Confession*. He was followed in this by a large number of Lutherans who came to refer to themselves as "Phillippists" (from his first name, Phillipp). In contrast, the "Flacians" (also "Exclusive Lutherans") insisted on Luther's *Augustana vera* and his concept of receiving the Lord's Supper. The vast majority of the Bohemian and a significant part of the Moravian Utraquists, who during the second quarter of the sixteenth century embraced the European Reformation—began to refer to themselves as "evangelicals under both kinds" and Hrejsa (1912) retrospectively referred to them as "New Utraquists" —and leaned towards the Phillippist version of Lutheranism.

From the point of view of the theory of the disenchantment of the world according to Weber, it is worth noting that in the *Philippist Confession* there is a distinct separation of the vehicle ("the first which is seen") and the signified ("the second which is not seen but heard of through the word of God"). This is characteristic of the Reformation's symbolic conception of the sacraments. It was most striking in Zwinglianism and Calvinism, but is also characteristic of the Philippists. The Unity of the Brethren also worked its way towards such a conception in the decades around 1600, a time of considerable convergence with Calvinism. The Brethren's Confession (originally dated 1535) says in its 1574 redaction, "And though both faithful and unfaithful (read: believers and unbelievers alike) receive the true sacrament and its truth sacredly and bodily, yet they themselves are faithful spiritually and so with a saving power." The 1607 redaction reformulates this passage, "And though both receive the true sacrament bodily (in the margin: by true sacrament we mean the material, elemental and earthly things, which are bread and wine), yet they themselves, believing in Christ, receive the truth of the sacrament (in marginalia: by the true sacrament we understand the spiritual and heavenly things, which are the body and blood of the Lord) spiritually through faith and so savingly" (Hrejsa, 1912, p. 314).

In the 1662 Amsterdam edition of the Confessions of the Brethren, Comenius returned to the original 1535 wording, and thus to the 1574 formulation of this passage. The "Sacred" can be received with the mouth; see above. Thus, in the 1607 text which leans toward a purely symbolic conception of the sacrament, the editors would presumably have been concerned with the notion that the "truth of the sacrament" could be transmitted and manipulated in any "sacred" (i.e., animate, in the form of bread and wine) way. For them, the "truth of the sacrament" is only disseminatable by the spiritual way of faith.

Similarly, in the Second Helvetic Confession of 1568, the signifier (bread and wine) is sharply separated from the signified (the truth or matter of the sacrament; *veritas* or *res sacramenti* (https://heidelblog.net/2013/10/the–second–helvetic–confession/). The signifying can be bodily grasped and ingested with the mouth. The signified can only be received by way of faith through the action of the Holy Spirit. The article on the Lord's Supper, for example, says: "And this is visibly represented by this sacrament outwardly through the ministers, and, as it were, presented to outer eyes to be seen, which is invisibly wrought by the Holy Spirit inwardly in the soul." The cut between vehicle and signified is so sharp that it emphasizes the inadmissibility of the assumption that the body and blood of the Lord are substantially present in the "elements of the sacrament" (i.e., in the bread and wine) by priestly transubstantiation: this would be to endorse cannibalism. "For as the flesh of Christ cannot be eaten corporeally without infamy and savagery, so it is not food for the stomach," follows a little further on in the Helvetic Confession (Říčan, 1951, p. 246). Thus, in a Reformed (especially Philippist and Calvinist) world, whoever would want to take "substance" in the Lord's Supper—to return to the above quote from Barfield—"into themselves, together with its representations, bread and wine"—would be suspected of cannibalism.

If one can only receive the "substance", the "truth" of the sacrament by faith—by the power of the Holy Spirit's grace—and if the mere manipulation of the bread (or the host) and the wine (or the cup) has no redeeming effect, it then appears that we are indeed in a totally de–magicized, rationalized and disenchanted world. It seems that we can identify the Reformation critique of the Catholic, supposedly magical conception of the sacrament of the Lord's Supper as a "culturally evolutionary event" within a historical sequence that is part of a line of development described by means—using Koselleck's words—as a "theory of possible history" of the "process of disenchantment of the world". Of course, this "theory of possible history" needs to be exposed to the possible "veto of the sources" (Koselleck, 2004, p. 151; Vašíček, 1994). And alas, the explicitness of the specific "theory of possible history" of the process of the world's disenchantment may be partially vetoed by the same source texts on which the theory relies. For the "elements of the sacrament" (bread and wine) are not disenchanted to all consequences, they are not completely without magical, enchanting power. We see this in paragraph 9 of Article XXI of the Helvetic Confession, where it says: "But he who comes to this sacred Table of the Lord without faith, communicates only in the sacrament and does not receive the substance of the sacrament whence comes life and salvation; and such men unworthily eat of the Lord's Table. Whoever eats the bread or drinks the cup of the Lord in an unworthy manner will be guilty of the body and blood of the Lord, and eats and drinks judgment upon himself (I Cor 11: 26–29). For when they do not approach with true faith, they dishonor the death of Christ, and therefore eat and drink condemnation to themselves."

These words could certainly be interpreted in terms of a moral lesson: specifically, that it is not appropriate to do without faith what others in the congregation are doing in faith. (But this would probably be an interpretation that would bring Enlightenment rationality to mental representations of the sixteenth century.) The possibility of reading this paragraph in this way could be reinforced by the text of the very next paragraph 10. It says, among other things, "The body of Christ is in heaven at the right hand of the Father; and therefore, our hearts are to be lifted up on high, and not to be fixed on the bread, neither is the Lord to be worshipped in the bread." Nevertheless, let us try to read the paragraph 9 of Article XXI in such a way that the "elements of the sacrament", the bread and wine on the Lord's table, can have a mysteriously magical damning power if they are "unworthily" handled. This element of the radical Protestant doctrine of salvation is also pointed out by Max Weber as one of the significant factors that led to modern self–conscious introspection. Under the conditions of a strictly predestinarian doctrine and the rejection of the confessional sacrament, it is necessary to question myself, lest I accidentally receive the sacrament without inner purification and without true faith. It would not be mere indecency towards others, it would be my fall into nothingness.

Thus, even after the breakthrough brought about by the Reformation, the "disenchanted world" contains a certain ambivalence (possible sources of "re–enchantment"). For example, František Žilka, originally an evangelical pastor and later a religious scholar, pointed this out in the middle of the twentieth century: "I think and see that the religious thinking and feeling of today's protestantism in the matter of the sacrament is on the whole along the lines of the consequences of Zwingli's conception" (Žilka, 1941, p. 233). The inclination to Zwingli's conception, that is, to the conception most radically disenchanting the sacrament of the Lord's Supper, Žilka further confirms with the words: "It can only be a symbol of satiation without magic. I believe that this conception is a necessary next step in the line of the Reformation" (Žilka, 1941, pp. 233–234). However, Žilka points out the tension between theological doctrine and lay thinking, even within evangelical congregations: "Most of the lay participants did not and do not have any distinct ideas about the dogmatic core of the controversy and positive views about the nature and form of the presence of Christ, but accepted or assumed the magical or mystical power and action of the sacrament. The proof of this from ancient times is the custom, not general, but nevertheless widespread, of administering the Holy Communion or the Lord's Supper to children, perhaps by analogy with the order of infant baptism which had long before been established. A more recent and lasting indication of this concept is the incorrect custom among some evangelicals of administering the Lord's Supper to the sick almost as if it were also a 'provisioning' sacrament at the threshold of death" (Žilka, 1941, p. 232). He has no choice but to conclude: "The sacraments are symbols, visible signs of spiritual values. Especially free Protestantism consistently removes from the sacraments everything magical that prevails in Roman Catholic dogma and to a certain extent is also maintained in the Protestant faith, as I indicated when I mentioned the ministry to the sick" (Žilka, 1941, p. 233).

This somewhat lenghty passage gives a very important lesson that can be found in many other communities (Closures 2). All Christians recognize the canonical text of the Scriptures as sacrosanct, yet they may quarrel concerning their interpretation up to the point of total discrepancy and breakup.

5.5 Plasticity and Elasticity in the Ontic or in the Noetic Sense?

The ambivalence mentioned above, whereby the world of Calvinistic reformation is both (mostly) disenchanted and (secondarily) re–enchanted, may be a source of the possible reversibility of the process of disenchantment of the world (Horský, 2009, pp. 182–205 and 294–305). The historiography of (popular) culture in the early modern and modern history, moreover, points out that this is not a process that would unequivocally and forever transform the Western world. As noted above, examples of manifestations of the "re–enchantment" of the world have been shown (Horský, 2009, pp. 167–173 and 186–196). The process of "disenchantment of the world" exhibits a (prevailing) plasticity.[2] It appears to be so especially with regard to the rationalizing–secularizing Enlightenment break, after which the disenchantment also entered Catholic areas (Tinková, 2004, pp. 157–232). At the same time, however, the "disenchantment of the world" can also be understood as a process of cultural and historical change, in which we can demonstrate a certain evolutionary elasticity. This lies in the tension between the demagicizing and spiritualizing efforts of the Reformation on the one hand and the magical inclinations of a part of popular culture, potentiated by elements of the Catholic Baroque, on the other. We must bear in mind, however, that as culturally evolutionary plastic changes, Enlightenment reforms (as for example in Austria, the imperial *Tolerantion patent*) appear to us from our present perspective. It cannot be ruled out that in two or three centuries' time even these changes may appear to be merely the manifestation of an elastic evolutionary fluctuation.

From a certain point in the development of culture, binary oppositions emerged in the field of religious "cultural content" (i.e., a sense of pure spirituality versus idolatry observed as early as the Old Testament prophets; and as observed in the Reformation of the sixteenth century, the sacrament of the Lord's Supper as a

[2] Evolutionary plasticity and elasticity are discussed in more details in Chap. 11. In short, elasticity points at trends towards conservative, established norms; plasticity refers to introductions of novelty (explosions).

symbolic reminder of Christ's sacrifice versus the sacrament of the Lord's Supper as one of the means of salvation.

The disintegration of spiritual content into such oppositional categories would, in Cassirer's perspective, correspond more to evolutionary *plasticity*. It would fulfill the definition of evolution or development with which the historical sciences work (Wittkau–Horgby, 2002) understanding it as an "irreversible, progressive, gradual change over time". Beyond that, it would also be an "increase in complexity" or "increase in differentiation" as the definition of evolution commonly used in biology would require. In contrast, *elasticity* would be exhibited (over a period of time) by the preferences of the relevant "cultural variants" or "contents" (either pure spirituality or idolatry; or the symbolic conception of the sacraments or their saving nature) as they would vary among the members of a particular culture (in their "mental representations") with their temporal, social, local or generational variability.

But are "plasticity" and "elasticity" categories historical, *ontic* (i.e., culture has indeed developed and is developing in this way) or *noetic*, methodological (i.e., these categories are our tools for interpreting the history of culture)? In the sense of a kind of intellectual game, we can ask if what we have here at the level of culture is analogous to genes and their alleles (as considered by Flegr, 2010, 2021; Toman & Flegr, 2018) within the theory of frozen evolution. Following Rickert, we can distinguish "cultural contents" or "cultural values" as "good, "beautiful", "true" and "holy" (each with possible opposites: evil etc.). Let us say that "cultural values" are, in a sense, *abstract* invariants and that their specific historical forms are their *concrete* variants. Do they allow an analogy with alleles? In both cases, such a relationship may be a source of possible evolutionary elasticity. We can regard them as constituted and differentiated within the (evolutionarily plastic) "process of the gradual illumination of spirit" and their worship being demagicized in the process of the separation of "existence" and "meaning". But today they can also be treated methodologically as abstract values, which leads our cultural–historical research to an attempt to thematize (retrospectively) their historically variable concretizations (Meran, 1988, p. 127). The process Cassirer has in mind takes place over time, from the birth of human culture and spirituality to the present. The second, which could be derived from Rickert's considerations, is retrospective. In this retrospective, the four values are not clarified by the historical process but are clarified by us *through our asking questions* and finding out the relation of their presumed concrete forms to the ideal value types of "goodness", "beauty", "truth", and "holiness".

If we are guided by the idea of a process of gradual disenchantment and spiritualizing of the image of the world, it can happen (and probably will happen) that we will consider one and the same factual aspect of a certain group of "traces" as an indicator of "(immediate) objects" (objects of research) that differ from those we would consider if we are guided, for example, by the idea that the disenchantment or, on the contrary, the magical image of the world is conditioned, for example, ethnically. Moreover, there is the question of the number of traces. The more that are available, the more selection will have to be made. The criteria for this selection, whether explicit or intuitively implicit, are (often) not neutral in terms of the theories and narrative interpretations measured by them but are (covertly) dependent on them. The foregoing may explain why Weber's and Barfield's assessments of the Reformation critique of the sacraments as a cultural–historical "event" differ, as will be shown below.

5.6 Weber and Barfield's "After"

For Barfield (1988), what came into being through the disenchantment of the sacraments is a *subjective void*, a zero point between original and final participation, to use his terminology (final participation, in simple terms, means that we ourselves take charge of the world). With reference to Koselleck, we can say that this is an example of how the same plot and content, the same configuration of traces or "incidence" can become a different "event" in the perspective of different authors. The nature and evaluation of this event will depend, in the conception of any given author, on the "before" and "after" that will be at the beginnings and endings of the sequences of historical lines in which these authors place the respective "incidence" when they construct it as an "event". The stretches between the "before" and the "after" in this case are quite long, spanning several centuries.

Max Weber and Owen Barfield would probably agree quite a bit on what that "before" is for them (the medieval conception of the Eucharist). But they differ on the "after". Weber (1991) understands modern science as a vocation. This science is one dimension of the very specific modern "formal rationality" of the West.

In Barfield's conception (1988), the emergence of modern science is also related to a certain analogy of the "disenchantment of the world" (though he does not call it that). But modern (scientific) rationality, in his view, leads to idolatry. Barfield's historical analysis, however, is not as value neutral as Weber's. Barfield attributes a certain finality to history. The notion of the aforementioned "zero point" on humanity's journey from original to final participation is reminiscent of "alienation" —as described in the nineteenth century classics of German philosophy. In this conception, entities that are immanent to development can undergo this "alienation" in their transformation from an initial state of a "thing–in–itself" (*an sich*; a state that is internally indifferent, dark, confused, unconscious) to a final state "for themselves" (*für sich*; a state that is fully internally differentiated, clear, distinct, (self conscious). "Thus, the possibility to look back over the history of the world and achieve a full, waking picture of his own gradual emergence from original participation, really only arose for man with the culmination of idolatry in the nineteenth century" says Barfield (1988, p. 183). In moving toward final participation, both the dogmatically conceived host and the chalice on the altar as well as the presumed elements of "objective reality" were all revealed as idols. A gradually historically acquired mindfulness also implies a capacity to understand religious texts as parables (Barfield, 1988, pp. 180–181). History, evolution, is thus conceived by Barfield as actually (quasi)religious:

> When the evolution of phenomena is substituted for our supposed evolution of idols, it will, I believe, be seen without much difficulty that the evolution of the individual human spirit has always proceeded step by step with the evolution of nature; and that both are indeed 'fallen'. The biological evolution of the human race is, in fact, only one half of the story; the other has still to be told. [...] The awakened clarity of retrospect [...] will, I am persuaded, be obliged to recognize that the gradual emergence of man from original participation amounts also to the gradual emergence of 'men' from 'man'; that it is not just the cumulative history of the race, but the biography, also, of each individual spirit (Barfield, 1988, p. 184).

We could also say that Weber and Barfield structure their "space of experience" differently because each treats the "horizon of expectation" differently. The category of the "space of experience" (*Erfahrungsraum*) is elaborated in detail by Reinhart Koselleck (2004, pp. 255–279) in a dialectical relation to the category of the "horizon of expectation" (*Erwartungshorizont*). "History" in his conception is not a continuous process in and of itself. The continuity of history is always and only *created* by people through their actions. People act, according to Koselleck, with respect to how they thematize their future, with respect to the "horizon of expectation", and they do so within the framework of their "space of experience", within the framework of accumulated historical experiences. However, these experiences are not just mechanically sedimented, but always, at each time, *newly structured* (as the reader may have already understood, we try to extend this insight to the history of the entire biosphere at various points in our text). These are experiences of all kinds (political, social, religious, scientific, environmental, etc.). The structuring of the "space of experience" is closely dependent on what is on the "horizon of expectation": each individual self in each of its Here and Now acts with a certain knowledge of the past and with responsibility for the future. These claims are easily transferable to all inhabitants of the biosphere. Admittedly, the reader must not equate the "horizon of expectation" with the teleological notion of a fixed, preconceived goal, a point on the horizon, these passages in no way point to a similar interpretation. History is both "given" and "open" on the ontic plane and on the noetic plane. This conception of Koselleck is followed by Ricoeur (1988).

The differently structured "space of experience" allows, for example, a completely different interpretation of commonly known texts or artefacts. An example from the history of the Reformation might be the "reading" of verses 53–56 of the sixth chapter of John's Gospel (J6: 53–56):

> Then Jesus said unto them, Verily, verily, I say unto you, Except ye eat the flesh of the Son of man, and drink his blood, ye have no life in you. Whoso eateth my flesh and drinketh my blood hath everlasting life, and I will raise him up at the last day. For my true flesh is meat indeed, and my blood is drink indeed He that eateth my flesh, and drinketh my blood, dwelleth in me, and I in him.

In the early 1420s, the Czech Hussites base their demand in the second article of the Four Articles of Prague on these verses: "that the glorious sacrament of the body and blood of God, under the both kinds of bread and wine, should be freely given to all faithful Christians, to whom mortal sin does not hinder, according to the ordinances and commandments of the Savior." For "the Savior binds all Christians to receive this sacrament under the death of the soul, which is a mortal sin." What follows is a reference to those verses (Říčan, 1951, pp. 40–41), which were read literally. Moreover, the emphasis is on communion in the sacramental plane (with the mouth). It follows, then, that all (i.e., even laymen and children) who are to be saved must receive communion. From the Second Article of Prague this also came into the Basel Compacts. The "space of experience" of the Hussites is thus imbued with a distinctly magical aspect. Scripture (at least in relation to the passage quoted) is read literally, the sacrament is perceived materially.

A hundred years later, in the 1520s, the teaching of Huldrych Zwingli on the Lord's Supper begins to penetrate the Bohemian lands. One of the first writings that contributed to this is known as the "Odpis" (Opposition) of Jan Čížek. Following the example of Zwingli, he also based his concept of the Lord's Supper on the text of the sixth chapter of John's Gospel. And like Zwingli, he reads it quite differently from the Hussites of a centory before him. Čížek interprets the text of this chapter, including the verses quoted above, as allegory and metaphor. His "space of experience" is imbued with a spiritualistic aspect.

These individual and situational "incidents" can then be considered "evolutionary events" within the process of the world's disenchantment, during which the differentiation and diversification of individual lines of development took place. Some of these have had a classical bifurcating form, others rather branch in many directions. In both cases, we can show mutually diffusing influence and sometimes even backward interconnection.

Luther posting the 95 Theses in Wittenberg in the autumn of 1517, the split of the Anabaptists from Zwingli's congregation in Zurich in 1525, and Calvin's return to Geneva can all stand as significant "symbolic centers" within images of history, which take on different connotations depending on who shapes this or that image of history. But are these "events" also real—whether objectively determinable or only culturally relative "historical centers" of "evolutionary events"? If they were "historical centers", then the question arises: to what extent would there be hidden contingency in them? Would it be true that if Zwingli had not written the treatise on the Lord's Supper, the "world" would have remained "without disenchantment"? Or could we assume that the tendency towards radical disenchantment was present in the "spirit of the age" such that if Zwingli had not translated it into explicit form, someone else would have done so (perhaps one of those who became radical followers of Zwingli during the 1520s)? This question cannot be decided argumentatively on the ground of historical scholarship. Any eventual answer will depend on our initial value positions, which are often only implicitly held. It is worth inquiring as to whether the situation of historians differs on this point from that of biologists when they try to assess the contingency of a particular macroevolutionary event, as previously described, in considering the relationship between "evolution" and contingency–containing "history".

But we must ask another question: to what extent and in what sense is the "disenchantment of the world" an actual "fact"? And if it is a fact, what have we proved by it? Have we factualized the disenchantment of the world? But even if we had, then it is only in the form of a certain reconfiguration of meanings and values within a semiotic and axiological system. Moreover, we have "proven" it only by taking certain textual, iconographic and statistical source traces as indicators that certain cultural contents (meanings, values) were actively present in the culture from which they originated. But can we consider this reconfiguration of meanings and values as a symptom of a change in the—shall we say—"psychic structures" of the people who lived at that time? If so, would it be a temporary, reversible change in their minds, or a historical change, i.e., persisting at least for several generations? We must be careful with our conclusions. There is always, to speak with Barfield, a danger of idolatry, and not only in science and history.

References

Angehrn, E. (2004). Kultur und Geschichte – Historizität der Kultur und kulturelles Gedächnis [History of culture & cultural memory]. In F. Jaeger & B. Liebsch (Eds.), *Handbuch der Kulturwissenschaften. Band 1: Grundlagen und Schlüsselbegriffe* (pp. 385–400). Metzler.

Ankersmit, F. R. (1994). Six theses on narrativist philosophy of history. In *History and tropology. The rise and fall of metaphor* (pp. 33–43). University of California Press.

Barfield, O. (1988). *[1957] Saving the appearances. A study in idolatry*. Wesleyan University Press.

Blickle, P. (1992). *Die Reformation im Reich*. Ulmer.

Burke, P. (1990). Historians, Anthropologists, and Symbols. In E. Ohnuki–Tierney (Ed.), *Through Times: Anthropological Approaches* (pp. 268–288). Stanford University Press.

Cassirer, E. (1946). *The myth of the state*. Yale University Press.

Cassirer, E. (1955). *The philosophy of symbolic forms. Vol. 2: Mythical thought*. Yale University Press.

Cassirer, E. (2020). *The philosophy of symbolic forms. Vol. 1: Language*. Taylor & Francis.

Flegr, J. (2010). Elastic, not plastic species: Frozen plasticity theory and the origin of adaptive evolution in sexually reproducing organisms. *Biology Direct, 5*, 2.

Flegr, J. (2021). *Frozen evolution. A farewell to selfish gene*. Andrias Press.

Geertz, C. (1977). *The interpretation of cultures*. Basic Books.

Ginzburg, C. (1989). [1980] Clues: Roots of an evidential paradigm. In *Clues, myths, and the historical method* (pp. 87–113). Johns Hopkins University Press.

Hanák, J. (1928–1929). Bratři a starší z Hory lilecké. *Čas Matice Moravské, 52*, 39–124 and 277–348; *53*, 1–44.

Havelka, M. (1999). Symbol – centrum – typus. Tři perspektivy řešení vztahu případu a zobecnění ve vědách o člověku [Symbol—center—typus. Three perspectives of generalization in human sciences]. *Acta Univ Carolinae, Philosophica et historica 4, Studia philosophica, XIV*, 24ý–245.

Havelka, M. (2012). "Odkouzlení" versus sekularizace? [Disenchantment or secularization?]. *Sociológia, 44*, 564–578.

Horský, J. (2009). *Dějepisectví mezi vědou a vyprávěním. Úvahy o povaze, postupech a mezích historické vědy. [Historiography between science and narration]*. Argo.

Hrejsa, F. (1912). *Česká konfese, její vznik, podstata a dějiny [The Czech confession]*. Česká akademie císaře Františka Josefa pro vědy, slovesnost a umění.

Huizinga, J. (1996). *[1919] The autumn of the Middle Ages*. University of Chicago Press.

Koselleck, R. (2004). *[1979] Futures past: on the semantics of historical time*. Columbia University Press.

Krakauer, D. C., et al. (Eds.). (2017). *History, Big History, & Metahistory*. Santa Fe Imstitute Press.

Kuukkanen, J. M. (2015). Why we need to move form truth–functionality to performativity in historiography. *History and Theory, 54*, 226–248.

Kybal, V. (1905). *M. Matěj z Janova. Jeho život, spisy a učení [Matthew from Janov]*. M. Marek.

Meran, J. (1988). Historische Methode oder Methoden in der Historie? Eine Frage im Lichte der Methodologiegeschichte. In C. Meier & J. Rüsen (Eds.), *Historische Metode*. Deutscher Taschenbuch Verlag.
Mink, L. O. (1970). History and fiction as modes of comprehension. In *Historical understanding, Ithaca*. Cornell University Press.
Monod, J. (1972). *[1970] Chance and necessity. An essay on the natural philosophy of modern biology*. W. Collins Sons & Co.
Müller, H. P., & Sigmund, S. (Eds.). (2014). *Max Weber Handbuch. Leben – Werk – Wirkung*. J. B. Metzler.
Odložilík, O. (1923). Jednota bratří Habrovanských [Assosiation of the Brethren from Habrova]. *Čes čas historický, 29*, 1–70. a 301–357.
Říčan, R. (Ed.). (1951). *Čtyři vyznání: vyznání Augsburské, Bratrské, Helvetské a České: Se 4 vyznáními staré církve a se Čtyřmi články pražskými. [Four denominations]*. Komenského evangelická fakulta bohoslovecká.
Richerson, P. J., & Boyd, R. (2006). *Not by genes alone: how culture transformed human evolution*. University of Chicago Press.
Ricoeur, P. (1984). *Time and narrative I*. University of Chicago Press.
Ricoeur, P. (1988). *Time and narrative III*. University of Chicago Press.
Rüsen, J. (2006). *Meaning and representation in history*. Berghahn Books.
Šmahel, F. (2002). *Die Hussitsche Revolution*. Harrassowitz Verlag.
Šmahel, F. (2006). *Die Prager Universität Im Mittelalter; Charles University in the Middle Ages (bilingua)*. Brill Acadamic Pub.
Spier, F. (2017). Complexity in Big History. In D. C. Krakauer, J. Lewis Gaddis, & K. Pomeranz (Eds.), *History, big history, & metahistory* (pp. 195–223). SFI Pres.
Tamm, M., & Simon, Z. B. (2020). More–than–human history. philosophy of history at the time of the Anthropocene. In J.-M. Kuukkanen (Ed.), *philosophy of history twenty-first-century perspectives* (pp. 198–215). Bloonebery.
Tinková, D. (2004). *Hřích, zločin, šílenství v čase odkouzlování světa [Sin, crime, and madness in te time of disenchantment of the world]*. Argo.
Toman, J., & Flegr, J. (2018). Macroevolutionary freezing and the Janussian nature of evolvability. Is the evolution (of profound biological novelty) going to end? *Biosemiotics, 11*, 1–23.
Třeštík, D. (2003). Dějiny ve věku nejistot [History in the age of uncertainty]. In Klápště et al. (Eds.), *Dějiny ve věku nejistot. Sborník k příležitosti 70. narozenin Dušana Třeštíka* (pp. 23–44). Nakladatelství Lidové noviny.
van Dülmen, R. (1987). *Reformation als Revolution. Soziale Bewegung und religiöser Radikalismus in der deutschen Reformation [Reformation as revolution]*. Fischer Verlag.
Vašíček, Z. (1994). *L'archéologie, l'histoire, le passé: chapitres sur la présentation, l'épistémologie et l'ontologie du temps*. Kron.
Weber, M. (1904). Die "Objektivität" sozialwissenschaftlicher und sozialpolitischer Erkenntnis. *Arch Sozialwissenschaft und Sozialpolitik, 19*, 22–87.
Weber, M. (1991). In M. Sukale (Ed.), *Schriften zur Wissenschaftslehre*. Reclam.
Weber, M. (2005). *[1930]. The Protestant Ethic and the Spirit of Capitalism*. Routlege.
Wehler, H. U. (1979). Anwendung von Theorien in der Geschichtswissenschaft. In J. Kocka & T. Nipperdey (Eds.), *Theorie und Erzählung in der Geschichte* (pp. 17–40). Deutscher Taschenbuch.
Welskopp, T. (2008). Theorien in der Geschichtswissenschaft. In G. Budde et al. (Eds.), *Geschichte. Studien, Wissenschaft, Beruf* (pp. 138–157). Akademie Verlag.
Wittkau–Horgby, A. (2002). Evolution. In S. Jordan (Ed.), *Lexikon Geschichtswissenschaft. Hundert Grundbegriffe* (pp. 88–90). Reclam.
Zeleňák, E. (2015). Two versions of a constructivist view of historical work. *History and Theory, 54*, 209–225.
Žilka, F. (1941). *Naše křesťanství [Our Christianity]*. Laichter.

Chapter 6
Non-contemporary Contemporaneity

Abstract We understand the non-contemporal contemporaneity in lines of Koselleck: as a situation when we distinguish—in a single socio–cultural field—the presence and impacts of various cultural elements of different date of provenience and of various dynamics of their transformation. We present here some modalities of such configurations, before all in the relation to narrative theories of historiography and the historism (as different to historicism). On the Czech history around 1900 we ask, if we can *ontically* assess a cultural integrity of an epoch. We follow the Wittlich's interpretation of the Štursa's sculpture *Melancholic girl* as a real cultural meta-attractor that unifies what can be called "Czech Art nouveau". We argue that such a unity is more likely but a result of a narrative historical interpretation (Ankersmit). This conclusion we also project into the concept of quasi–character (Ricoeur, Mandelbaum) that we do not link to whole societies, but only to plenty of interpretative communities. These arose in different times and with a different temporality of their transformation, and can coexist or even overlap in a single socio–cultural field, all representing cases of Closure 2. Hence, we see analogies between the evolution of biological species and of socio–cultural complexes, We return again to the topics of plasticity and elasticity of (cultural) evolution. Non–temporal contemporality might also somatize—in course of human evolution.

Keywords Historism and historicism · Cultural variant and cultural complex · Quasi–character · Somatization of cultural evolution

> For the history lying in the heads of "ordinary" people, what is fashionably called "memory" among professionals, is far from identical with what historians write in their textbooks. [...] It is a junkyard on which the broken fragments of history are strewn; and at the same time, it is not a junkyard, because it is alive. People relate to it, use the discarded fragments of history secondarily as a tool to help them make sense of their own world. [...] Collective

A. Markoš, J. Horský, *The Seeds of Forms and the Forms of Seeds*, Biosemiotics 12, https://doi.org/10.1007/978-3-032-16131-4_6

> myth-making memory is not the recollection of some vague and actually unknowable past, but it is the recollection of what is said about it.
>
> (Dušan Třeštík, 2008, pp. 158–159)

Carlo Ginzburg (1989) draws attention to the forensic methods of historians "interconnecting history, archaeology, geology, astrophysics and paleontology", that is, the ability to pronounce oracles retrospectively. On "an interpretative method based on debris, on marginal data, considered as detectors. This way, details usually considered unimportant or even trivial, 'low', could provide the key for understanding the highest product of human spirit" (Ginzburg, 1989, p. 280). Intuition, unpredictable elements, a single glance, scents, etc., play an important role in medicine, in police investigations, news reporting, expertness of arts, psychoanalysis, archaeology, historical sciences, etc. Thus, Ginzburg is particularly inspiring for us here because his conception of the paradigm of clues strongly supports the possibility of thinking through analogies between history and evolutionary biology. At the same time, however, Ginzburg's example also serves by showing the diversity of contemporary considerations within the theory and methodology of the historical sciences.

6.1 Contemporary, Non-contemporary, Narrated

The role of historiography, according to Ankersmit, as we already know, is essentially interpretative: "to find unity in diversity". According to narrativist critiques, historiography brings this unity to diversity through linguistic, literary operations such as plots or narrative sentences.

A plot is defined by Paul Ricoeur as a *juxtaposition of the incongruous* or the *configuration of the heterogeneous*. By devising a plot, the historians integrate such events into a connection for which they have neither empirical (source) nor theoretical reasons. Ricoeur (referring to Cassirer and Geertz) considers "narrative structure" and "plot" as symbolic tools that allow us to make an experience of the passage of time (through narrative structures) and the experience of both our actions and the actions of others (Ricoeur, 1984, pp. 88–128).

Arthur Danto (1965) considers narrative sentences by which *historians* (e.g., not *witnesses* of an event, who do not have the appropriate temporal distance, or are often no longer among us) narratively *construct* the historical (meaning) *coherence* of events. A narrative sentence (said in the present) is, for example: "The Protestant part of the Czech estates lost all hope of religious freedom in 1648." However, we could (and should) add to what Danto said: narrative sentences are constructed by historians but also by witnesses (actors) in terms of the experiencing of the event itself or within the lifelong experience of the passage of time. Among other things, these *explicit* but also *implicit* narrative sentences structure the experience of individual situations (in shorter periods of time) or the experience of their life-time (in longer periods of time). For example, "The player who came onto the ice from the bench missed me on the right wing" is a narrative sentence, just as "I sold the house my father built in 2004". These sentences meaningfully connect subsequent events in time (i.e., they constitute their historical connection), even though they are *not necessarily* causally related: the substitute player did not necessarily miss me; the sale of the house was not necessarily related to the fact that my father built it; yet from my perspective as a narrator, these events are *perceived* as *meaningfully related.*

The question, however, concerns the extent to which the *semantic connection* that is constructed in narrative sentences is also a connection that is—shall we say—socioculturally and historically *ontic* (or, in our thematizations, *ontological*). Answers to this may differ between the narrative sentences of a historian (as considered by Danto) and those of (what we assume here) the actor. So far as narrative sentences in the researcher's terms are concerned, they would not (in the spirit of narrativist historic nominalism) be ascribed ontological significance by most authors. However, when it comes to actor's narrative sentences, historical sciences are offered the opportunity to take inspiration from biosemiotics. The latter reckons with the "reality" of semiotic relations (Deely, 2008, 2010). Processes of meaning attribution, i.e., processes of semiosis, are real, ontic processes (Švorcová & Kleisner, 2018) and are of constitutive importance in the evolution of life. The same could probably—at least from a certain point of view—be said of the meaning contexts that are constituted in an actor's narrative sentences.

"Contemporaneity" is, from the point of view of the theory of historical sciences, a much more complex category than the use of the word in ordinary language might suggest. Neither from the psychological point of view, nor from the point of view of the theory of history, is it some kind of immediate "now". In each moment, we experience a multitude of retentions (reverberations of what has gone before) and protensions (anticipations of what is to come; Mayerová & Vašíček, 2004, p. 11). We can only structure the "present" or "now" through our experience, insight into perceived "events" or "situation", at a specific point in time, always subsequent to and with the need for a certain (albeit sometimes very short) temporal distance.

Some retention reverberations can be very long. Many Berliners who experienced the building of the Berlin Wall may have seen its construction as a turning point ("event") that established their "present": but this lasted three decades. The Wall was "new" from their point of view because it was not "self-evident" perhaps even on the eve of its fall. Those who were born when the wall was already standing may have perceived it as "ahistorical" because it was "self-evident". It only became "historic" for them when it fell. "Past" can thus be distinguished from "present" only by considering those contents that are no longer part of (actual) "memory" as "past", i.e., those whose retention is not perceptible from an actor's point of view (Mayerová & Vašíček, 2004; Maslowski et al., 2015). Recollection, the recall of memory contents into actual consciousness, can take either intentional or merely implicit sociocultural form as processes of commemoration (Assmann, 2002), or it can be initiated, for example, merely by accidental sensory stimuli (Ginzburg, 1989). One can then distinguish not only "memory" from "history", but also "history" from "historicity" and "past" from the "former" (colored, endowed with meaning) by the historicity of the human "sojourn" (Ricoeur, 1985).

The "contemporaneity" is thus permeated by much of the "non-contemporary". Reflections on the non-contemporary have a long tradition (e.g., Vašíček, 1994). Paul Nolte (2002, p. 134) defines the contemporaneity of the non-contemporary as a theoretical formula designating the specifically modern experience of the calendarium contemporaneity present of historical structures and processes belonging to different developmental phases and layers of origin from the past. Nolte considers this contemporaneity of the non-contemporary primarily in relation to "big history": for example, after the discovery of the New World, Europe encountered cultures that were "non-contemporary" because they were at a different stage of development from Europe. But it is also possible—beyond Nolte's interpretation—to consider a smaller territorial scale of individual socio–cultural fields in the European environment. Moreover, contemporary non-contemporaneity can also be considered in a completely micro-analytical manner specific to individual groups, generations or individuals. A particular socio–cultural field can thus be viewed in terms of the simultaneous correlation of many entities that differ in the time of their emergence and the temporality of their transformation (see Fig. 11.1). Thus, we can observe "non-contemporary contemporaneity" in at least three ways.

First, at the level of actor's experience, we can note "the difference of generations living at the same time, each having different experiences, relating to the past differently, and thus living in a shifted phase" (see above the example of the Berlin Wall). The same "incidence" can be perceived individually, collectively, or generationally from different actor's perspectives as part of different subjectively experienced "horizons of the present". Thus, the time span whose content we are trying to meaningfully integrate as "now" ("present") by means of plotting can be individually, collectively or generationally distinct.

Second, we can simultaneously consider the presence of different cultural layers in the same "contemporary" socio–cultural space, despite being "non-contemporary" in origin. We must consider that these layers carry their immanent meanings with them, without having to be currently present in the actor's consciousness. In the 1980s, for example, a dispute emerged as to whether a ritual—even one of very ancient origin—constitutes meaning without the actors being aware of it. Darnton (1985, 2009) and Medick (1984) considered that it did; Chartier (1985), on the other hand, emphasized the need to take actor consciousness into account. Thus, we consider the "non-contemporaneity" of different semiotic, axiological and praxeological systems of culture. In later chapters, especially 11, we will draw attention to similar analogies of overlapping time horizons in the evolution of communities or lineages of other living creatures; in both cases, the plastic and elastic elements of the evolution of communities are mixed. At this point, however, let us continue with an analysis of the history of human cultures.

Thus, in a one time-plane of a sociocultural field, patterns of culture, cultural complexes and variants, working practices, patterns of behavior, interpretations, rhetorical patterns, etc. of different ages and different dynamics of change are all present. In their interpenetration and competition, questions arise as to what to do today with ancient cultural values: what to do, for example, with the core of the gospel message in modern times (Bultmann, 2019)? Can we use new architectural and artistic forms (e.g., churches and church interiors) to reimagine sacred spaces that historically had few established forms?

Third, we can build on this and think of "non-contemporary contemporaneity" of objects in terms of things, texts, artifacts, even e-memes (inactive, external memes). For example, texts may turn into i-memes (active, internal memes in people's minds) after a certain period of time (see *hypolepsis*, 2.3, 4.1, 11.2, and *passim*). However, we must not proceed hastily; the whole mechanism of these new actions of old objects and texts is more complex than can be described as the mere proliferation of memes. We might see here an analogy with how organisms handle their genetic information and experience.

In human history, the accelerating increase of the "non-contemporary" that correlates with each other in one-time horizon (i.e., the possible increase of heterogeneity in the socio–cultural field) can retroactively dynamize further change. This can be shown with the example of "fashion" in the sense that Georg Simmel (2004) thinks of it (as well as of "fashion" in "inventing" history in order to "illuminate" contemporary events). It is related to the extent to which homogeneity is demanded in any given culture (or, on the contrary, heterogeneity is tolerated) and to the extent to which change is considered legitimate in that culture. Let us now consider variation, heredity, and selection as the three principles constituting evolution, since they represent the basic evolutionary principles. "Fashion" in Simmel's conception is thus closely related to "cultural evolution". The more variable the fashion (including "non-contemporary" elements) there is in a particular socio–cultural field, the more likely the dynamics of cultural change will be in that field.

6.2 Historism and the Historicization of Ethics

During the time that the coexistence of the "non-contemporary" was increasing in the socio–cultural field, historism was simultaneously born and became, in the opinion of many authors, the defining thought form of the nineteenth century (Iggers, 1997, p. 459). It significantly shaped twentieth century thought, and in many ways is remains active today. "Historism" is, however, a highly ambiguous term; moreover, in many languages there is an added distinction between "historism" and "historicism".

Historicism relies on the existence of a law of historical development, on the recognizability of this law, and on the possibility of a future subordination of ethics and behavior to this law. This historicism, especially in the thought of Hegel and Marx, was criticized by Popper (1988). He rejects Hegel's (also applied by Marx) "philosophy of identity" as a distinctive feature of historicism in which the "real" (what is historically established by development) is also "reasonable" and therefore "right". Some proponents of historicism (e.g., Ernst Troeltsch or Max Weber) would agree with Popper, who, like Michael Ruse (2017), would distinguish between a thing as a result of evolution and what it is normatively seen to be.

Historism developed in parallel to historicism. It was closely related to the constitution of a specific "historical method" intending to differentiate what is (being; Sein) from how it happens to have been historically constituted (formation; das Werden; Meran, 1988). Developmental tendencies can be recognized ex post, but nothing binding for the future can be deduced from them. At the same time, historism combines a developmental perspective with an attempt to explain the phenomena under study from their contemporary context. With this, however, came the question of whether and to what extent ethical, aesthetic, religious, or cultural values can be historicized (i.e., possibly relativized).

Similarly with the notion of "progress": even if we could, for example, demonstrate increasing differentiation and hence increasing internal intellectual richness (which Weber does not rule out), we could not decide (i.e., evaluate) on the ground of "empirical science" whether some "progress" has taken place. (This does not mean, however, that the "ethically right" cannot somehow be subordinated to historical, cultural, or social circumstances.)

With regard to questions of the historicization of ethics, Ernst Troeltsch (2024) distinguishes three levels: first, the level of the "general spirit" or "spirit of the time": this is actor-unconscious, and historically and culturally relative. The second is the level of "cultural values": this is creative personalities that have the ability to express certain tendencies of the "spirit" of their "time", to make them period-conscious. This gives rise to "cultural values" that are explicit, consciously pursued ideals at the time, values that "should apply". There is thus a time-relative ethics of cultural values. Thirdly and finally, Troeltsch discusses the level of a personal "ethics of attitude", which is timeless, ahistorical. It should be added, however, that already some of his contemporaries, e.g., the theorists of the so-called "axial age" and "axial transformation" (Alfred Weber, Karl Jaspers), would argue that this plane, too, was established by historical development. (Biologists more or less think outside of similar categories, but a kind of unacknowledged historicism can be found in evolutionary biology, probably only on a figurative level.)

Historicism has made it possible to take a (partially) relativistic stance on many phenomena from the past. We can thus say of them that they are "former", i.e., that we would no longer agree with them "now", yet we must accept them as they "once" happened. In this diachronic perspective, the boundary between "once" and "now" is variable. For example, the holocaust is still "now" for "us" in terms of values (i.e., its historical relativization would probably be considered a threat to the values that apply "now"). On the other hand, the Franco–German hostilities of both World Wars now seem to be mutually seen as "past". Yet, at the same time, for some people, for example, the burning of Jan Hus is "now", i.e., Catholicism is still to blame for it. Historians, however, have long viewed it as "former".

6.3 Cultural Integrity of an Epoch: An Example of the Period Around 1900

Let us choose an example of a time horizon around 1900. We offer the reader examples from Czech cultural history at the end of the "long 19th century" (as historians often refer to the period from 1789 to 1914–1918). Around 1900 we find both novelty in many social phenomena and a striking development of knowledge in many disciplines, as well as a widespread belief in one's own modernity and one's own contribution to progress. At the same time, however, we can also observe an unprecedented diversity of cultural expressions. We can either attempt to historically unify this diversity, or on the contrary, use it to trace the interplay of cultural complexes in the socio–cultural field.

To generate historiographic unifying tools, for example in Ankersmit's conception, to *represent* cultural diversity by means of a narrative historical interpretation with the greatest "scope" in relation to the diversity of individual statements that objectively refer to individual cultural phenomena, will be the most metaphorical of all narrative interpretations that—in our case, in relation to "Czech culture" in the time horizon around 1900:

"A historical narrative is a historical narrative only insofar as the (metaphorical) meaning of the historical narrative in its totality transcends the (literal) meaning of the sum of its individual statements. Being a historical narrative, therefore, is a matter of degree" (Ankersmit, 1994, p. 41, thesis 5.2.1). In fact, Ankersmit envisages three such degrees: first—historical narrative interpretations may be named "Names like *Mannerism* refer to historical interpretations and not to past reality itself" (Ankersmit, 1994, p. 39, thesis 4.7.1); second—however "the name Mannerism refers to the statements of a narrative interpretation"; and third—"in these statements, reference is made to historical reality" (Ankersmit, 1994, p. 39, thesis 4.7.3).

Let us at this point leave aside the fact that historians such as Chartier (1996), Ginzburg (2013), and Koselleck (2004, pp. 150–151) would probably disagree with Ankersmit's confidence in the unproblematic objectivity of factual "statements". Indeed, they advocate a softer theory of fact (Lorenz, 2022). We are here looking to Ankersmit for support for the claim that the totality of the picture of a (cultural, historical) epoch offered by historiography very likely has no ontic counterpart in the form of a real structured interconnectedness of all the cultural phenomena of a given time.

We are therefore inclined to the idea that in any given socio–cultural field, in any given time horizon, there are different, mutually competing cultural complexes which may differ in the date of their origin and the temporality of their transformation. Let us begin with the example of an interpretation of the Czech Art Nouveau offered 40 years ago by the Czech art historian Petr Wittlich (1982). Wittlich drew attention to the considerable cultural and historical significance of the period. However, he did not consider it merely one of several artistic movements of the turn of the 19th and 20th centuries but presented it as an expression of the more general cultural climate of the era.

Wittlich paid considerable attention to Jan Štursa's sculpture Melancholy Girl from 1906 (Fig. 6.1). He says of this work: "It was already recognized by its contemporaries as a true contemporary symbol, summarizing the poetic concept of youth as the central theme of Czech Art Nouveau. A first sketch of the Melancholy Girl has survived, which was probably made as a record of a sudden idea, as it was drawn over smaller sketches of funerary monuments. [...] In the bowed young body of the seated girl, whose face resembles more of a mask, and in the curve of her arms around her head, Štursa has found a dynamic spatial ornament that fully captures the absolute sense of the Art Nouveau curve. In the ambivalence of perfect form and the immediacy of feeling, in the emotional transcendence of the artifact, Štursa's *Melancholy Girl* became perhaps the most stylish example of Art Nouveau proper in Czech sculpture" (Wittlich, 1982, p. 268).

Wittlich's way of evaluating the Melancholy girl makes it possible to partially revise Ankersmit's (1994) narrativist theory. First, it turns out that "historical narrative interpretations" integrate not only partial factual statements that are in the nature of *references*, but also such partial statements that are themselves partial *metaphorical representations.* Moreover, Wittlich offers a masterful demonstration of how the whole (the "narrative substance", called "Czech Art Nouveau") can be defined by *pointing to a single artifact* which—from the perspective of Wittlich's

Fig. 6.1 Jan Štursa (about 1906) *Melancholy girl.* Courtesy Moravian gallery Brno

interpretation—has the *capacity to represent* this whole *metaphorically*. Štursa's *Melancholy Girl* is "a real contemporary symbol" (Wittlich, 1982, p. 268).

However, this metaphorical representation of Art Nouveau contains a certain internal contradiction that becomes apparent if by Art Nouveau we do not mean an artistic movement, but an entire epoch. For Wittlich, the term "(dynamic spatial) ornament" is significant (Wittlich, 1982, p. 268). He notes the contemporary use of this term by the Czech impressionist painter Antonín Slavíček, who remarks (in 1908) about his perception:

"One observes and forgets everything. We see neither forms nor lines—and see only a kind of phantoms of colors—they bend closer to us—they gather together; at that moment the tree has neither branches nor leaves and has only fiery spots—in some—never before known ornaments" (Wittlich, 1982, p. 229). After this quote reverberates, Wittlich adds: "Slavíček's statement is only from 1908, but it is his last word that shows well how strongly the basic Art Nouveau ties were working. The ornament was still understood by Slavíček as the overall sign and ground of the picture—only it was an ornament hitherto unknown, or, more precisely, hitherto unread by the artist, but existing in itself in the book of nature" (Wittlich, 1982, p. 229).

Slavíček's use of the term "ornament" may be merely a manifestation of a discursive practice that was perhaps symptomatic of the time, but not intended by the author. However, even if Slavíček were to associate the term with a particular theoretical concept, it would be questionable whether it could be interpreted in terms of Art Nouveau aesthetics. Ornaments can be found in the paintings of Gustav Klimt or Alfons Mucha and can be seen in the decoration of much Art Nouveau architecture. But are they really a constitutive element of the structure of Slavíček's impressionist paintings? Wittlich thinks of "dynamic spatial ornament, fully capturing the absolute sense of the infinite Art Nouveau line" in connection with Štursa (Wittlich in Lahoda et al., 1998, p. 316). But can we associate Slavíček's "ornament" with the "infinite Art Nouveau line" when Slavíček explicitly says that the painter does not experience either lines or forms?

"Ornament" is one of the interpretative tools Wittlich uses to capture ex post the spirit of the (Czech) Art Nouveau as an epoch. Ankersmit, within the framework of his strict narrativist nominalism, would posit that this category functions as a means of linking and unifying interpretation and as a vehicle of metaphorical representation only in the plane of the historian's text, but that it had no real ontic counterpart in the past. However, this nominalist distinction need not be respected. Let us say that here we have a certain feature ("ornament") that is either directly empirically visible on a plethora of artifacts or, with a certain amount of interpretative imagination, observable. We can therefore declare it to be a symptom of the unreflected "spirit of the age" (epoch). However, we must be cautious, (a) not to immediately and without further ado confuse the *symptom* with a *formative principle*, (b) nor to regard it as a principle deliberately applied by the *author*, and (c) not to succumb to the impression that this principle—if it existed at all on the ontic plane—was the *exclusive* formative force. In the same empirically graspable source material, multiple other unifying features are likely to be found, depending on the initial criteria of observation and the aesthetic values chosen.

Štursa's girl is melancholic. This is in line with the attentiveness and sensitivity towards (often highly individualized) subjectivity and innerness that was present in the socio–cultural field during the timeframe from around 1900. This permeates both contemporary noetic thought and aesthetic or psychological considerations. In particle physics, "matter has disappeared", the indisputability of the impenetrable solid of Newton's world has passed into the relativized certainty of the phenomenal world of Einstein and Planck. Analogously, in painting, the "real object" has "disappeared", as we can read in Slavíček's statement quoted above. Both the visual and literary arts themselves and their contemporary critics provide numerous examples of subjectivizing (some even naturalizing) psychologization on the horizon of the "one-time present" (Welskopp, 2008) around 1900. The legitimacy of considering sensitivity to subjectivity and to innerness as a characteristic of the times is supported by the development of religious feeling and thought.

According to Ernst Troeltsch, the (Christian) religion of modern humans is spiritualistic mysticism, since it promotes autonomously individualistic sentiments and is a bulwark against clericalism (Troeltsch, 2024). Yet also, religion has had to address within itself the question of how to deal with traditional, ancient cultural contents or cultural values that stood out with their claims to validity. Purely as a working example, let us choose two biblical passages. They relate to the "kingdom of God", and to wha it means to be "born–again":

> And when it was demanded of the Pharisees, when the kingdom of God should come, he answered them and said, The kingdom of God cometh not with observation: Neither shall they say, Lo here! or lo there! for, behold, the kingdom of God is within you. And he said unto the disciples, "These days will come, when ye shall desire to see one of the days of the Son of man, and ye shall not see it" (Luc 17, 20-22).
>
> and
>
> Verily, verily, I say unto thee, except the man be born again, cannot see the kingdom of God [...] I say unto thee, you must be born again. (Joh 3, 3-7 [both quotes from King James Bible]).

What to do with these Bible verses and these concepts in a time of cultural, social and political modernization? "Kingdom of God" or "born again" certainly sounded a lot from both Catholic and Evangelical pulpits around 1900. Outside of churches and places of worship, however, in the wider socio–cultural field, these terms could already sound quite archaic and "out of time".

Let us now move to the mountain village of Křížlice in Krkonoše (Giant) Mountains, the highest mountain range on the North Bohemia. Evangelicals were a minority in the region, compared to the Catholic majority, but they were not insignificant in Křížlice, which had about 700 inhabitants. The pastor of the congregation, František Hrejsa, notes a decline in the number of members around 1900. The reason for this seems to be the folk spiritism that spread in Křížlice and its surroundings and which permeated the Evangelical congregation as well (Kletvík, 2017).

Křížlice is sometimes cited as the place where Czech spiritualism was born around 1880. The first two known spiritist mediums (both women) in the region came from Křížlice. Now, the biblical concept of "born–again" was used in sermons by the pastor Hrejsa and shared by members of his congregation, but at the same time it was also used by spiritists. However, each interpreted it differently. The temporal and cultural variation in the interpretation of "born–again" will be of interest to us. From a theological point of view, the tensions between (evangelical) Christianity and Spiritism are undeniable and obvious. From a cultural-historical or cultural-anthropological point of view (Troeltsch, 1931), it can be said that *spiritism* is congenial onto-mystical *spiritualism*, and the spiritist idea of an ongoing *new birth*.

The tradition of Reformation knows a twofold basic concept of "born–again". The first is in the sense of being born again by faith every time we confess our sins in our hearts. According to Luther, if we have faith then by the power of that faith and the word of God within us, we become born again. The second is the concept of a new birth as a personal restructuring that is a turning point in the life of the individual, which was also developed during the Protestant Reformation.

The concept of "born–again" as a personal transformation began to emerge quite early in the European Reformation. Its roots can be seen, for example, in the mid–1520s, when the Anabaptists (Täufer) separated from Zwingli's congregation in Zurich and, in a consistent assumption of a mystical–spiritualistic, symbolic conception of the sacraments, began to apply baptism only in adulthood. The external baptism in water was for them only a symbol of the inner baptism or the baptism of the Spirit. This inner baptism was then merged for them with the inward acceptance of Christ (with his mystical transubstantiation), i.e., with a personal transformation, a "new birth" (Dülmen 1987, p. 184). At the same time, the new birth appears in the reflections of some Lutheran and Calvinist authors (without any connection to Anabaptism) at the turn of the sixteenth and seventeenth centuries. Here, to be born–again is an almost cosmic event, an act by which the individual transforms his/her cosmic situatedness. Hieronymus Zanchius, Johann Arndt, and Johann Valentin Andreae, among others, associate "new birth" ("Wiedergeburt") with breaking free from the physical influence of the stars (Mahlmann, 2004; Repo, 2004).

During the seventeenth century, the concept of "new birth" subsequently passed into Lutheran pietism—probably also due to the influence of the mysticism of Jacob Böhme (Wehr, 1995, pp. 236–249). Pietism generally emphasizes intrinsic individualism, which is related to the idea that one belongs to a congregation (church) by one's own free decision. Lutheran Pietist congregations (as "little communities in [country] churches") are then "communities of the born–again" (Troeltsch, 1931). Lutheran Pietism, however, is in accord with Luther's doctrine of life within or as a secular "vocation" (Beruf). In Lutheran pietism, inner individualism is therefore of a purely religious nature but has no immediate relevance for human action in the (socio)political sphere. In some congregations in the eighteenth and nineteenth centuries, Lutheran pietism encouraged a self-reflexive methodical conduct of life. At the same time, it was sometimes permeated with an interest in natural philosophy and natural science (Gleixner, 2005).

Lutheran pietism (especially from Halle and Herrnhut) began to influence Czech (secret) non-Catholics during the eighteenth century. We see this, for example, influencing the pietistic congregation of the "church of grace" in Těšín, and permeating the Czech exile congregations in Lusatia (in Herrnhut), Saxony and Prussia (in Berlin). After the issue of the Toleration Patent of 1871, however, the question of "born–again" does not seem to have been important for those Protestants in the Czech lands who opted for one of the tolerated denominations (i.e., either Augsburg or Helvetic). The theme of "born–again" then reappears in the nineteenth and twentieth centuries in evangelical congregations in the Czech lands with various, but not strictly pietistic, accents.

However, it was different in the case of those (until then secret non–Catholics or Protestants) who during 1782 began to subscribe to non–tolerated, and often unknown, denominations, such as the "Faith of God" or the "Faith of the Lamb", etc. They often identified themselves confessionally by "breaking bread and drinking wine from the cup". At the same time, however, for many of these to be "born–again" probably carried more weight than it did for those who chose one of the tolerated denominations. Beyond that, these "errorists", as the contemporary state administration began to label them, often overemphasized "new birth", and understood it differently from Lutheran pietism. We have, for example, a relative abundance of records of interrogations resulting from persecution by the Habsburg state, wherein such individuals give their age in years that had passed since their "new birth", rather than their biological birth.

The contemporary Czech religious scholar and sociologist Zdeněk R. Nešpor (2003) distinguishes between "folk spiritism" and "urban spiritism". Perhaps to both can be applied the contemporary definition of spiritualism by Zachar (1914) as a movement that "believes that the human soul is immortal and is able to communicate with living people in this world even after the death of the body, and during the time of its appearance it even performs various physical and mental acts that a living person [...] is not capable of. In order that the living world may come into contact with the spirits of the departed, a medium is necessary." After citing this definition, Nešpor (2003, p. 65) adds that "in reality, however, spiritism was more a conglomeration of various ideas, coinciding in this way but differing in another, differing in place and time."

According to Nešpor, Czech folk spiritism grew out of the "errant religious ideas" and "folk sectarianism" of the eighteenth and nineteenth centuries (Nešpor, 2003). Because of this, it already contained numerous religious elements and had the character of a religious sect with its emphasis on ascetic life practices, inner (spiritual) piety and love for neighbors (Nešpor, 2003, pp. 65–71). Folk spiritism connects with folk sectarianism and Lutheran pietism through (a) an emphasis on the personal experience of a sudden awakening in faith by the action of the Spirit (Nešpor, 2003), which is in line with the notion of "born–again", and (b) the reading of Scripture. By reference to Scripture, some of the spiritists in the Giant Mountains and the surrounding region around 1900 argued directly in favor of their doctrine. The role of the medium was explained, among other things, by the biblical text, "I will pour out of my Spirit upon all flesh: and your sons and your daughters shall prophesy" (Ac 2:17).

In the horizon of the "former present" around 1900, intellectual evangelical circles treated the concept of "born–again" in connection with the "Kingdom of God" differently. They sought the intersection of ("contemporary") modernity and ("non-contemporary") religious cultural variants. Many shared with Troeltsch the conviction that the kingdom of God is within and among us. Troeltsch's "within and among" can be read in two ways: either, say, in a mystical-spiritualist way, namely that the kingdom of God is in our (individual) innerness (in the heart, in the mind, in the soul); or in a social-spiritualist altruistic way, namely, that it is within our congregations. Troeltsch's sympathy for mystical spiritualism would suggest the former. However, his notion of the sanctification of this world is more likely to mean the latter. Similar reflections on the kingdom of God can be found in František Žilka (1941, pp. 306–322), a prominent representative of Czech "free Protestantism". He considers three possible conceptions of the kingdom of God: either it is in heaven (i.e., in the future, or in the Church), or it is in the soul, or finally it is being actively implemented in the world (Žilka, 1941, pp. 310–311). He himself takes a kind of intermediate position between the second and the third answer to this question. "I believe that a better world is within reach", he says. If it can be attained, "why should it not be possible by the inner power of Christianity, by the power of brotherly love" (Žilka, 1941, p. 313). Žilka also offers us an overview of several competing interpretations that occurred in the timeframe around 1900.

The phrase "Except a man be born again, he cannot see the kingdom of God", which can be described as a certain discursive regularity, thus took on a variety of meanings in different times and in different socio–cultural settings, i.e., it could and did become one or another cultural variant. In doing so, it was integrated into various cultural contexts that were both diachronically and synchronically competent with each other. Some of these cultural complexes can and could be constitutive in the emergence of certain communities (cf. in the following subsection the correspondence between "quasi–characters" and "Closures 2").

Let us try to apply Weber's (1972) typology of asceticism along with Troeltsch' (1931) and Giddens' (1989, pp. 450–480) typological distinction between "churches" and "sects" to the above examples of different interpretations of "born–again" and "kingdom of God". All these interpretations, with a partial exception in the case of Anabaptism, accommodate "inner-worldly asceticism", i.e., they do not require an "extra-worldly-ascetic" escape from the world and exclusion from society. Although the born–again are in a specific cosmic or religious situation, they remain in the social bonds of this world. However, both the Lutheran-Pietist and the Spiritist model fulfill more of the features of a "sect" (Nešpor, 2003 considers the folk spiritists directly a "sect" in the sense of Troeltsch's typology). This can be said especially regarding the emphasis on the *ascetic purity* of the born–again or awakened. This purity then establishes the *exclusivity of the community* compared to the wider society in which it lives. In the timeframe circa 1900, this solution was no longer "contemporary", but nevertheless remained productive in constituting a certain minority.

Troeltsch and Žilka's way of *internalizing* the Kingdom of God and bringing it "into the world", i.e., projecting it into everyday *social interactions*, also requires an inner "new birth". However, this new birth is not conceived as dramatically as it is in the case of the more religiously exclusive societies of pietists and spiritualists. Troeltsch's and Žilka's conception of the domestication of the kingdom of God and its translation into the human world is in tune with the modern *privatization of religion and faith*, or the transformation of religion into (more modernly conceived) mysticism. In his typology, Troeltsch understands "mysticism" as an alternative to both "churches" and "sects". In the timeframe around 1900, it is a more "contemporary" solution than the pietistic or spiritualistic one. Moreover, it is a solution that can be applied throughout society.

We could certainly try—returning to Ankersmit's concept quoted above—to give a unifying interpretation of the whole culture of the Czech lands around 1900 by means of a plot within a coherent narrative interpretation. In doing so, we could try to choose the "Art Nouveau line" and "ornament" or "a psychologizing sensibility towards individual innerness" as a unifying interpretative tool. However, we feel that even if we succeeded, the most we could achieve is an interpretative unity that lacks an ontic counterpart. Therefore, we will respect the diversity of cultural phenomena in their eventual configuration. After all, alongside Art Nouveau and Impressionism, one can find manifestations of Symbolism, Expressionism, Cubism and abstraction in Czech art and architecture—all at the same time, all with the reverberations of historicizing pseudostyles. In the same way, alongside mystically attuned piety, we also find an abundance, if not a majority, of traditionally ecclesiastic forms. One could go on.

Examples of the competitiveness of cultural variants and cultural complexes within one-time frame reinforce the validity of the assumption that biological and cultural evolution are isomorphic. At the same time, in our opinion, it might be more productive to look for analogies other than that between the evolution of species and the evolution of cultures. Rather, it seems that the analogous counterpart of a biological species might be a cultural complex, with phenotypic (morphological or behavioral) traits corresponding to the individual cultural variants (as components of the cultural complex). The analogous counterpart of a cultural complex (or socio–cultural field) would then be what biology considers the environment and its pressures and influences.

6.4 Quasi–Character

We have already used the term quasi–character (*quasi-personnage*; Ricoeur, 1984) several times in our interpretation. It is time to define this term in more detail and to link it to other terms we use in this book (such as Closure, "symbolic center", "space of experience", "horizon of expectation", etc.).

History, as has been noted here many times, routinely writes about and studies large formations, wholes such as "society", "culture", "civilization", "nation", "church", etc. These are very often treated as "quasi–characters" in historical interpretations. They are spoken of, and probably even thought about and researched, as if they were individuals endowed with the capacity to think, feel, want or act. This way of writing has probably come into historiography from political speech, but it is common in ordinary language as well. The terms historians use to refer to "quasi–characters" are thus often taken from the language of the actors they study. Thus, it is not without significance for us to ask whether quasi–characters correspond "in reality" to an internally interconnected, relationally interdependent whole. For in the sociocultural field, according to many historians (e.g., Koselleck, 2004), the general concepts used by actors are not neutral. Actors think in these terms, and such terms enter into the motivations of their actions and help them to relate to each other. It is an open question to what extent the performativity of a particular concept, or, conversely, the extent to which the success of actors' attempts to use such concepts constitutively, depends on extra-linguistic factors and factors operating outside mental representations (e.g., the "national idea" and its "chance" of achieving "objectification"). By this extent we mean the capacity of a certain concept, term or word when uttered, to break into the socio–cultural field and to restructure this field, or at least certain segments of it, in a *constitutive* way. The Geschichtliche Grundbegriffe school of history, represented for example by Koselleck, advocates a perspective in which such concepts are both a factor of sociocultural structurizations but also their symptom, an instrument of successive articulations of what is already structured outside language and outside thought. In contrast, there are schools of Anglo–American historiography that tend to place great emphasis on the unmediated and direct performativity of concepts and ideas (Lottes, 2002). Such a *degree of performativity* cannot be decided unequivocally, as it certainly varies in different times and in different social and cultural contexts. There are situations that, for example, Weber (1972, pp. 1–30) as well as Müller and Sigmund (2014, pp. 58–63 and 90–92) refer to as *charismatic authority* (charismatische Herrschaft), in which the mere utterance of one who is recognized as the "lord" changes the sociocultural field. But at the same time, there are many other situations in the same society in which the same words that had performative force in the exceptional moment of charismatic authority come across as mere rhetorical exercises empty of content.

If a historian names a certain group as *they* (or as *we in the past*) or if some actors refer to themselves as *we*, i.e., if a certain quasi–character is created by historians or actors on the narrative level, this does not immediately and without further ado make it possible to answer unequivocally the above question as to whether, or to what extent, such a discursive quasi–character corresponds "in reality" to a whole socio–cultural formation (cf. above on the difference between ontic continuity and the narrative meaning of events). Let us name three of the possible responses as nominalist, essentialist and narrativist.

In the sociologically *nominalist* perspective, society is no more than a name, no more than the sum of interacting individuals. The historians' "they" or the actors' "we" (e.g., "Czechs", "peasantry", "Catholics") are merely conceptual tools that serve the social scientists' knowledge or actor's orientation. The essentialist approach presupposes that "they" or "we" —i.e., quasi–characters thematized on the linguistic level—can only correspond to socio–cultural formations if they refer to real "ideas" (in Hegel's sense) or real socio-economic structurations (in Marx's sense; also see above concerning historicism).

The narrativist solution requires more attention at this point. It would link the "quasi–characters" to the necessary linguistic structure of any narrative. A strictly narrativist position (e.g., Mink, 1970; Ankersmit, 1994, 2001) would attribute existence to these "wholes" only at the level of narrative language. It would therefore also be strictly nominalist.

With regard to our thinking about possible analogies between biological and historical-scientific thought (or analogies between the development of life and the development of human culture), a solution that links the narrative level of the historical text with the social–ontological level is significant. Paul Ricoeur (1984) says, that "society, when treated as a single entity [...] acts as a quasi–character in historical discourse". This, according to him, has two reasons. The first, which can be described as narratological, is that from the point of view of narrative theory the nature of a quasi–character is anything that is "presented as a grammatical subject in a particular narrative, equipped with the predicate to act". As shown in Lorenz (2022), the second reason is social–ontological and is related to Mandelbaum's theory of societal facts. In addition to psychological facts (the thinking and acting of individuals), there are also social facts (structures and institutions), and these (in their full extent) cannot be deduced simply from psychological facts. On the contrary, the properties of acting individuals, characters, can be transferred to social realities, even in a kind of ontic sense. This possibility is based, among other things, on the fact that any definition of social facts ultimately refers indirectly to individuals, and is reinforced, as Ricoeur points out, by the three features by which Mandelbaum defines society: namely its territorial organization, the structure of its institutions, and temporal continuity (Ricoeur, 1984).

However, Mandelbaum and Ricoeur's idea that socio–cultural reality can take the form of quasi–characters even at the ontic level requires a certain nominalism-inspired caution. This is particularly true when we consider entire modern "nations" or "churches" as such quasi–characters. The legitimacy of thinking of such large quasi–characters not only in linguistic but also in socio–ontic terms is reinforced, for example, by the closeness of this concept to Gellner's (2009) reflections on the socio–ontic role of "holistic concepts". Indeed, according to Lorenz, Ernest Gellner offers two basic "ontological arguments" in favor of methodological collectivism. Firstly, he points to the existence of statistical facts, and secondly, that there are "holistic concepts" (just like "nation", "civilization", "Protestantism", etc.) that are not just neutral classificatory tools of researchers or actors but enter into the organization and motivation of the social actions of individuals and groups. Gellner's "holistic concepts" arguably overlap with Ricoeur's "quasi–characters" on more than one occasion. This coincides with Miroslav Hroch's (e.g., 2018) notion that an ethnic group or a particular community *constitutes* itself as a modern nation by, among other things, *narratively constructing* itself (mostly in a historically conceived narrative) as a nation. However, under certain circumstances that are far from being of a linguistic-literary nature, "narrating oneself" as a "nation" can mean becoming such a "nation". But the same is true, for example, of a reformist religious movement, etc. But not every such "narrative" is socio–ontically constitutive (see above about the "chance" of a certain "national idea" achieving its "objectification").

However, if we attribute the nature of quasi–character to smaller formations, if we allow that multiple quasi–characters can occur simultaneously in one and the same territory, and if we allow that one and the same actor can participate in multiple quasi–characters, we very likely increase the legitimacy of thinking about quasi–characters in a socio–ontical sense. The notion of "quasi–character" can then be linked to the notion of Closure 2, as seen elsewhere in this book (in this context, see especially Chap. 11).

If we view quasi–characters as Closures 2, we can assume that those who belong to a particular quasi–character, insofar as they experience themselves—always from just a certain point of view and in just a certain context—as "we", then and to that extent they can exhibit a common innerness or experience a common "world".

Let us return to the question of the common experience of the world, which we understand in Barfield's (1988, p. 19) sense, as a shared "set of collective representations". We introduce here as a presupposition: those who share "the world" exhibit—presumably—a common "figuration", i.e., a "creative work" through which, "mere sensations must be combined and constructed by the percipient mind into the recognizable and nameable objects we call here "things" (Barfield, 1988, p. 24). Those who share a common "figuration" are close in the ways they sensorially, introspectively or socially cognitively perceive and experience the world.

We can then construct, in case of small, native communities, so-called "indigenous peoples", a model of a relatively closed "quasi–character" to which we might attribute a very specific form of "figuration". We would thus be following Barfield's reasoning when he quotes Lévy–Bruhl saying that "Primitives see with eyes like ours, but they do not perceive with the same minds" (Barfield, 1988, p. 30). A textbook example would be the Inuit's ability, described by Franz Boas, to distinguish many more shades of snow than Europeans. Let us leave aside the fact that such a textbook example, like most such examples, is long outdated in its claims. Rather, we use it as an example for the following—we stress that it is purely hypothetical—consideration: if the Inuit were indeed the bearers of such a specific "figuration" that would ground their peculiar capacity for sensory perception and awareness of differently colored snow as distinguishable "objects" (all in the sense of Barfield's terminology), we could say that these are a "quasi–character" that exhibits a unique and peculiar "innerness" or "Closure 2", a "world" of its own. We may then ask how they maintain this ability and see three possible answers. It could be inherited genetically in the form of some gene mutation (though there is no empirical reason to suppose such a thing). So too, it could be inherited epigenetically (though here too we have no empirical grounds to suggest such a thing, but as shown elsewhere (Chaps 10 and 11), the possibility of epigenetic inheritance deserves considerable attention). And finally, it could have been inherited by vertical (or even horizontal) cultural transmission as a cognitive cultural a priori, either in the form of a general symbolic system in their culture, or just in the linguistic layer of that culture.

The terms used here ("quasi–character", "figuration", "closure", "world") need to be treated nominalistically in relation to the history of (human) culture; they are tools we have found useful in interpretation and knowledge. If we establish a certain "facticity" with their help, we must be aware that this facticity is of a very soft nature. As we work with the idea of the existence of such quasi–characters and their properties in the social–ontical plane, we must assume that their interfaces are very fluid, and that in the lives of individual actors these quasi–characters are necessarily activated only in the contextual of specific situations.

6.5 Plasticity and Elasticity in the Development of Culture

Let us return to the question of plasticity and elasticity in cultural-historical development, which we previously discussed in Chap. 5. We also said above that we would try to consider the analogy of the evolution of biological species (and their phenotypic traits, both morphological and behavioral) with the evolution of cultural complexes (and the cultural variants they contained). Within one "culture" or within the whole socio–cultural field or the whole culture-space of an "epoch" there are not only, as just said, multiple "Closures 2", but also multiple cultural complexes. We have shown in the previous chapter that whether a situationally configured "incidence" will appear to us as a cultural-historical "event" within a plastic evolutionary process depends, among other things, on the temporal perspective in which we establish Koselleck's "after". Considering what has been said in this chapter about the "non-contemporary present", we can also add that plasticity or elasticity cannot be related to the "culture" of the whole "epoch", but to the individual cultural complexes ("phenotypes") occurring within it, or to the cultural variants contained therein. Each of these cultural complexes may, as we already know, exhibit a different temporality of its transformation (we will return to this topic in Chap. 11).

An example that brings us back before the 1800 time-horizon may be the argument applied in the "bell ringing to prevent thunderstorms" controversy, as the study by Himl (2010) shows. One Enlightenment reform intended to educate "uneducated people" was the prohibition of such ringing. Ringing church bells to prevent a thunderstorm was forbidden in the Habsburg monarchy by a patent of Emperor Joseph II on November 26, 1783. This reform, however, met with relatively significant disapproval and even opposition not only in rural but also in urban areas. The clash was a continuation of an earlier controversy over the same topic: according to complaints from some estates, during the 1770s "enlightened" teachers, whose duty it was to ring, either did not want to do it or rang "little" or "shortly", thus from the point of view of "the people", they failed to prevent damage caused by lightning and hail. In fact, the imperial patent was repeatedly issued in 1788, 1800 and 1803. What deserves our attention is the reasoning used in the dispute over the ringing.

The Enlightenment state, citing scientific knowledge, forbids ringing against storm clouds not only because it is superstitious and ineffective, but also because it is dangerous. Both scholarly writings of the time and the patents contain the argument that the metal bell on church steeples attracts lightning and that the likelihood of a lightning strike on a church steeple is increased by the movement of the bell. We can say, then, that there is a certain cultural complex of "Enlightenment scholarship" that already shows a rather considerable developmental plasticity before 1800. But there are also arguments in favor of ringing. They have both their ecclesiastical, let us say "religious", form, and their folk form, which we might describe as "magical". In the "religious" argumentation, it is said that the ringing of the bell is meant to urge the faithful—in addition to taking the necessary precautions, taking shelter, driving away livestock, etc.—to pray to God or the saints to protect them and their goods and crops from lightning and hail. According to the magical argumentation, bells that have been blessed can better spread prayers, spoken at their blessing or sometimes written on them, by their movement into the air. Prayers spread in this way—by the physical movement of the bells—can more easily counter the devil's vicious efforts. We could say that we have two mutually competing cultural complexes, the "religious" and the "magical". This competition of complexes (or variants within one complex) shows elasticity, in which either "disenchantment" or "(re)enchantment of the world" prevails alternately and contextually in time and in the socio–cultural field.

It is not impossible, however, that within a certain timeframe (e.g., around 1800) and through its plasticity, an emerging and competing cultural complex (e.g., "enlightened scientific and rational" …) develops and demonstrates a tendency to pull some other cultural complexes towards a broader cultural change. It could be the case that a Lotmanian cultural "explosion" could occur (see Chap. 4). However, this plastic tendency, brought into the wider socio–cultural field, can be dampened by an unexpected reversal. For example, the conflict between "educated rationality" on the one hand and the "superstitious" tendency of "uneducated folk" to apply magical practices on the other, which from an Enlightenment perspective might have appeared irreconcilably antagonistic, was mitigated in Romanticism by the ethnographization and historicization of folk tradition. Indeed, certain efforts to defend bell ringing against thunderstorms not as an effective measure but as a traditional custom can be found even shortly after the 1783 patent (Himl, 2010). It is in fact another way of "disenchantment" or "demagication". For the romantic educated classes, such practices can become valuable by being folk and historical. The dispute over whether they are functional or not is now a minor curiosity. A new source of culturally evolutionary elasticity has come into play: folk "magical" practices are preserved but relabeled in value terms.

Admittedly, from the perspective of evolutionary biology, this evaluation of the various forms and transformations of cultural complexes and cultural variants may appear too soft. Thus, the possibility of determining whether they are in a plastic or elastic phase of their development does not appear to be indisputable and unambiguous.

6.6 Can Cultural Evolution Somatize?

We come to highly controversial stories of human evolution that are supported by very "soft" (if any) sources. In Chap. 9, we report on the search for "humanization genes" and their possible correlation with brain development. Here we will give examples "from above"—of possible evidence of changes in human thought during the several thousand years of the "Neolithic Revolution", which we can reconstruct, albeit very imprecisely.

By granting the capacity for semiosis to all life, by considering the potential of epigenetic inheritance, and by introducing categories such as "Closure 2", biosemiotics is loosening the boundary between "nature" and "culture". This makes it all the easier to ask whether culture or its development might somehow be *inscribed* in the course of (the biological) evolution (of humans).[1] We can imagine many physiological, behavioral and cognitive traits that are both phenotypic features of (not only) humans but can also be quite closely linked to certain "cultural contents". These characteristics have certainly been established *somehow* by evolution. But the question is whether and to what extent within the framework of biological evolution or within the framework of cultural development (both the evolution and diffusion of culture) and whether and to what extent biological and cultural development have interpenetrated or influenced each other. We can work out at least six ways in which this evolution could have taken place:

(a) Only in the sense of Darwinian evolution.
(b) In the sense of the co-evolution of genes and culture. In this we include the processes of "double inheritance" or the inheritance of genes, cultures and environments (Richerson & Boyd, 2006). At the same time, for example, evolution can be described by the theory of human self-domestication, where "culture" as the source of selection for cooperativeness and non-aggression, would also fall under this point.
(c) Furthermore, cases of possible epigenetic fixation of some cultural variants (cultural contents) can be considered. Attention is drawn to the *phenotype-first* principle advocated by some evolutionary biologists (West–Eberhard, 2003; Švorcová & Kleisner, 2018). This assumes that phenotypic changes precede genotypic changes in evolution (and the question is whether this includes culturally influenced changes in human behavior and cognition). In doing so, new phenotypic traits (phenotypic acculturation) are not necessarily fixed genetically but may sometimes only undergo (temporary) epigenetic fixation (Švorcová, 2023).
(d) Some characteristics have been established by cultural development. In doing so, we can imagine a twofold possibility. The first can be thought of in the context of what is said in Chap. 9 in relation to Portmann's thesis on the premature

[1] In other words, we ask whether culture has the power to influence either or both the genetic script and the settling of usage (e.g., in ontogenesis, metabolism, regulation life cycles, or in erecting new structures).

fetus. If humans differ from other great apes in that they are born "prematurely" and that their brains do not "mature" until the postnatal stages of individual ontogeny, this could mean that social learning (culture) is already significantly more influential in shaping cognitive and behavioral traits in humans than it is in apes or other animals.

(e) Another possibility is the establishment of relevant properties also by cultural development, taking place at a deeper level and on a broader scale (in Ricoeur's terminology, at the level of the "texture" of action) than the development of language ("texts" of action), but without any influence on immediate postnatal ontogeny. This would be, for example, the sense of Geertz's conception of culture as an "external program" that enables the human brain to function as it does (Geertz, 1993). Seen in this way, a particular cultural variant is inherited (only, or also) culturally.

(f) Finally, one can also consider the possibility that the above-mentioned properties are evolutionarily established (only) at the level of (cultural) language development and are inherited only culturally, by social learning.

The differing degrees of pervasiveness of biological and cultural evolution can be hypothesized and controversially considered in various examples. Below we give three: discussions of the so-called "grandmother hypothesis"; the linguistic (cultural) conditioning of color perception; and the "civilizing process". Finally, we mention the ideas of "somatization" of such cultural conveniences. Note how in all these cases there is a mixing of biological and cultural-sociological hypotheses. Note again, however, that these are ideas, hypotheses and narratives based on very "soft"—too soft—sources or data.

6.6.1 The Grandmother Hypothesis

Among other things, this hypothesis assumes that female menopause was evolutionarily established in women because it became advantageous from the point of view of children's *fitness* (their chances of future reproductive success) for their (maternal) grandmothers to care for them, especially in their infancy and young adulthood (inclusive *fitness* theory; Hawkes, 2003). Thus, it is assumed that what was originally inherited (only) by cultural transfer (the cultural "institution" of the grandmother) has passed into the plane of genetic inheritance (the evolutionary provisions of menopause). Thus, an ancient evolutionary event is assumed, the symptoms of which we are trying to find by the demographic analysis of some early modern (seventeenth to nineteenth century) local populations; in some, the "grandmother effect" can indeed be found. This consists in the fact that—compared to the rest of the population in question—children who are also cared for by their maternal grandmothers show lower infant mortality, and their mothers have shorter inter–natal intervals (Fox et al., 2017; Havlíček et al., 2021). Exceptionally, a positive effect of maternal grandmothers and a negative (infant mortality-increasing) effect of paternal grandmothers can be documented for a given local population (Voland & Beise, 2005). Cultural phenomena (demographic behavior) may serve as a hypothetical indicator of past evolutionary biological changes in the sense of (a) above. However, this does not necessarily mean that cultural evolution played a role in this change, let alone that it did so in the spirit of the above assumption. If so, it would presumably only be in the sense of the linkage described in (b). To assume an influence of culture on biological development at the onset of menopause in the sense of (c) above would probably be a very bold hypothesis. The other alternative, of course, is epigenetic inheritance: just as the institution of grandmothers had been established, it may again fade out after some number of generations.

6.6.2 Color Vision

There are many studies that focus on the correlation between language and color perception. For example, following Boas' cultural and linguistic relativism Lenneberg shows that monolingual native speakers of *Zuni* do not discriminate between and do not show competence in cognitive processes capable of perceiving the differences between yellow and orange, whereas bilingual individuals (Zuni with English) show a tendency towards such competence (Lenneberg & Roberts, 1956). Even linguistic universalists (Berlin & Kay, 1969; Kay & McDaniel, 1978; Kay & Maffi, 2000) know that within their theory of *basic color terms* members of some cultures have difficulties in distinguishing between green and blue, and that there are no differentiated terms for these colors in their languages. At the same time, we can also ask whether a native English speaker, if he or she were to learn the language of the Inuit (although this example is now considered more of a false story, but let us proceed with the analysis nonetheless), would also begin to "see" different colors of snow as they do. If so, then the ability to process such distinctions cognitively would be carried only by language, in the sense of (f) above. Or would he acquire this capacity only if he learned the Inuit language in infancy (when the brain is still "setting")? In this case the ability to discriminate colors would be based on the impact of language (or broader culture) on mental structures. This could correspond to (d) or (e) above. We could also imagine that such an impingement on cognitive structures (in our case, using the Inuit as a chosen example) was inherited not only culturally but also organically. Then we could consider the processes above in (c). Just to demonstrate the possible reasoning, we could also consider that the ability to differentially perceive snow colors has a genetic basis in the Inuit. If this were the case, some variant of the theory of self-domestication would hypothetically be offered to explain such a fact. The influence of environment and subsistence led to the selection of individuals for the ability to discriminate snow of different qualities (and thus colors) as finely as possible. This would then fall under (b) above.

The influence of the environment and alimentation could also be manifested at a more "prosaic" level. Most mammals develop an inability to digest fresh milk as they age (lactose intolerance): the production of the digestive enzyme lactase gradually declines, and in older individuals milk consumption leads to severe digestive problems. However, in many human populations the enzyme is maintained throughout life, and they can drink milk. Today, this is a mutation in the regulatory system, but the question remains as to whether, under nutritional pressure, some livestock cultures first developed long-lasting epimutations that were only later assimilated genetically. Other pastoral cultures, where such a shift in regulation did not occur in time, have changed their dietary habits to consume fermented milk: *this* established habit no longer requires evolution towards altered lactase regulation and intolerance persists.

In his theory of the civilizing process—as mentioned above in several places—Norbert Elias assumes a gradual historical stabilization and strengthening of the unconsciously functioning apparatus of human self-control. Elias describes similar (assumed) phenomena (facts), as does the theory of self-domestication (e.g., the historically increasing capacity of people to control their aggression). However, Elias considers this civilizing process across a much shorter time span (mere centuries—the decisive break according to him is the period between 1500 to 1800), rather than the entire time span in which the self-domesticating transformation of man is posited to have taken place (across thousands of years, up to ten thousand). Thus, the evolutionarily emergent trait of self-domestication, the increasing capacity for the internal differentiation of id and ego (or *superego;* Elias, 2000), presumably cannot (yet) assert itself at the levels of culturally influenced gene and allele preferences (see (b) above). The question, however, is whether this quality is only inherited socially and culturally both during the postnatal "maturation" of the brain (d) and later (e), or also by some organic route (c).

6.6.3 *Somatization*

We are approaching now a very sensitive and perhaps socially dangerous point. The subject of this book, however, can hardly avoid it. Recognizing that it is an entailment of mostly "soft" data and the vivid narratives built thereupon, we submit the following with some hesitation. The crux of our upcoming dilemma lies in questions of the permeability of the barriers between our two kinds of closures. Until recently, there was a straightforwardly apodictic belief that (genetic) information could flow only from the "eternal" Closure 1 to various forms of Closures 2. Thus, if a property of some Closure 2 is inherited transgenerationally, there could be only two "mechanisms" of such inheritance: "blind" execution of instructions from Closure 1 (admittedly, given the environmental context); or learning, whereby the new generation acquires certain properties only with the help of instruction from the previous generation (i.e., only if such an overlap of generations exists).[2] We emphasize: while the "aptitudes" encoded at Closure 1 persist, even if for some reason they

[2] Note that phenomena like genetic assimilation (see Chap. 10) have not permeated into the field.

do not necessarily manifest themselves in the following generation or generations, the experience acquired by an individual or group of individuals must have the possibility of non-genetic transmission within the group (i.e., through enduring Closure 2). For the human being we will discuss below, this possibility is extremely important: language, narrative, ancient writings, morality, and established cultural customs hold human communities (of different levels) together, but a young child adopted by a different group holds none of this and becomes a "renegade". The assumption of the heritability of (lifetime) acquired traits was until recently a heresy, indeed a taboo[3] in biology and culture: for good reasons that perhaps need not be elaborated here. Phenomena such as transgenerational epigenetic inheritance or genetic assimilation, however, have opened Pandora's box, reviving earlier speculations by historians and psychologists. So let us approach the problem of somatization not from a biological perspective but from that of culture; in other words, let us ask whether environmental pressures can to some degree permanently alter the thinking of large human groups or civilizations.

We are referring to the notion of a kind of collective mind ascribed to Closure 2, which society undoubtedly is. "Collective minds are problematic because there seems to be nowhere for any collective mind to reside (other than in the separate minds of the individual members of the collectivity). Here we encounter a more general difficulty about attributing views or characteristics to a community, and in particular a large community" (Harris, 2009, p. 5). What exactly is the force that shapes the mind of an individual? Harris (2009, p. 6) quotes Durkheim (2013): "I am not obliged to speak French with my fellow–countrymen nor to use the legal currency, but I cannot possibly do otherwise. If I tried to escape this necessity, my attempt would fail miserably."

The changing nature of the great civilizations over the last few millennia has aroused the interest of many scholars, and there is no shortage of hypotheses as to the causes of the onset of the various phases. In connection with the subject of our book, we report below on three such breaks, well recognized by many authors, which may have their correlate in the structure of the brain, specifically in the change in the relative importance and interconnection of the brain centers. These are the bicameral mind hypothesis; the influence of writing; and the mass adoption of digital technologies over the past few decades. We will not, therefore, deal with the theory of the so-called Axial Age, often cited in this context, according to which in the first millennium BCE (800–200) there was a relatively synchronous change of thought, and transition towards new types of monotheistic religion in the four most important civilizations of that time. The theory of the bicameral mind discussed below is close to these considerations, but it covers a much longer period, and more importantly, it attempts to relate the transition to changes in brain structure. We note

[3] *Nature* essayist Erika Hayden (2013) analyses four other kinds of such taboos that are better left unspoken: intelligence, race, violence and sexuality. Perhaps we should add (5) drug addiction, including common pharmaceuticals, and (6) raising human population. For example, S. J. Gould (1996) analyses the issue to a much greater extent (and with a historical dimension).

in advance that the depiction is tailored to Mediterranean civilizations using alphabetic writing—and to us, their heirs.

The Bicameral Mind Originally a psychologist and psychiatrist, Julian Jaynes opened his book (1990) by listing some facts about human behavior. He notes that thinking does not have to be *conscious*; consciousness is an extension of the ability to navigate in the world as other living creatures do. In contrast, humans do think and speak, yet it is striking that the earliest written sources seem to show that human thought was devoid of metaphors, the basic constituents of languages today (see, e.g., Lakoff & Johnson, 1980). The synthesis of these two claims then results in a speculative picture of humans as lacking subjectivity (consciousness of the self), yet understanding and obeying commands. Thus, they behaved like puppets fulfilling without reflection the commands, the "voices"—of whom? Of the gods? This hypothetical state of thought lasted, as far as we can very unreliably estimate, from at least the Neolithic Revolution 10,000 years ago (but perhaps as early as 40,000 years ago), and it began to break down during the second millennium BC.

Let us now turn our attention to the structure of the human brain. Anatomically, the two hemispheres (or "chambers" —hence the *bicameral* mind) of the forebrain are almost identical and are connected by massive bundles of nerve fibers (commissures). Functionally, they control the opposite sides of the body and are mostly equivalent and largely interchangeable. An important exception to this rule is the three speech centers; in right-handed people these are concentrated in the left hemisphere, and although not anatomically distinct, those in the right hemisphere perform the functions imperfectly and only after long training. This functional asymmetry has been detected by studying various injuries and lesions of the brain; by irritating the brain centers and then recording their electrical activity; by measuring this activity during various tasks (speaking, reading, reasoning, following commands, etc.); and after selective "sleeping" of one of the hemispheres, etc. Schizophrenia, epilepsy and other states of consciousness, as well as various injuries or lesions caused, for example, by tumors, have also served as sources of information.

On the basis of these scientific findings as well as on the basis of historical evidence, Jaynes hypothesized that people of archaic times were dominated by the right hemisphere, which produced "voices", i.e., hallucinatory sensations perceived as speech (probably sung or in poetic declamation). One could not oppose these commands, but obeyed them unconditionally.[4] This is evidenced by records dating back to the gradual transformation of man into a being endowed with consciousness. He finds evidence for his theory beginning with the Assyrian Code of

[4] Perhaps the current division of *today's* thinking into 'fast and slow' (Kahneman, 2012) gives us an imperfect picture of the situation. While today's person does not hear "voices", he or she automatically performs many routine activities "the way they are supposed to be done". At the same time, he or she may then engage other aspects of thought to perform unusual tasks (e.g., philosophical conversation during routine car driving). In Jaynes's interpretation, it is as if that "slow" thinking were missing or in an inferior position.

Chammurabi and the ancient Indian records; he has, of course, gained more certainty in the records from the beginnings of Mediterranean civilizations, although these records were often written down several centuries after the events depicted, and were subsequently edited many times. The gradual transformation of Greek thought after the catastrophe caused by the collapse of Mycenaean culture (Iliad, Odyssey, Hesiod), with its many quotations, illustrates the picture. From the same general period (the first half of the first millennium BCE) we get the stories that were layered into the Old Testament. The earliest prophets (e.g., Amos) routinely communicate with the Lord who appears to them. Later, the Lord only speaks (Samuel), and increasingly sporadically (Kings Saul and David), until he disappears altogether (Ecclesiastes). Just a minor illustration, the book is full of them: "Amos is almost pure bicameral speech, heard by an illiterate desert herdsman, and dictated to a scribe. In Ecclesiastes, in contrast, God is rarely mentioned, let alone ever speaking to its educated author. [...] In Amos there are no words for mind or think or feel or understand or anything similar whatever. Amos never ponders anything in his heart; he can't; he would not know what it meant. [...] Angry forceful speech which he probably does not understand himself. [...] Ecclesiastes thinks, considers, is constantly comparing one thing and another, and making brilliant metaphors as he does so. Amos uses external divination, Ecclesiastes never. Amos is fiercely righteous, absolutely assured, nobly rude, speaking a blustering god-speech with the unconscious rhetoric of an Achilles or a Chammurabi. Ecclesiastes would be an excellent fireside friend, mellow, kindly, concerned, hesitant, surveying all of life in a way that would have been impossible for Amos" (Jaynes, 1990, p. 296).

Similarly, the Psalms or David's songs express regret that man no longer hears God's voice and must help himself, on his own responsibility; and yet the texts lament in rememberance that only a few generations ago the Lord dealt with people personally: "Why have you forgotten me? Why must I go about mourning, oppressed by the enemy?" (Psalm 42:9) Further development of this kind of spiritual orientation continues only through minority groups: prophets, oracles, shamans, witches and similar people, who usually say something in trance that they do not remember in waking state, and the commands must therefore be interpreted by listeners. Often it ends by expulsion and even murder of such "obsessed" (we also find such incidents in the Bible); this continues today, when such people are locked up in asylums.

We do not have space here to report on Jaynes's remarkable book in detail, nor do we know the further development of the topic since the book was first published in 1976; here we stick to the context of the epigenetic processes discussed below and their possible heritable fixation. The brain is a plastic organ, and under the pressures of a changing civilizational "ecosystem", and it may well have rearranged its connections very radically, thus severely limiting the right hemisphere in its ability to influence speech centers and thought. Perhaps writing could have helped a lot.

The Literate Society If we return to the Durkheimian "collective mind" mentioned above, we can assert that a literate society is one in which a writing system has become the accepted practice of the majority population. As we saw in Chap. 4, writing represents a reduction of language which also allows for an *explosion* into new areas of world–making, including the emergence of rationality. However, this whole process carries with it the danger of limiting interpretation *to the point of view* of the literate mind (i.e., from *our* point of view). It may seem to us that an illiterate community is not rational, and it is with this premise that we can then evaluate the societies that Jaynes—as we saw above—calls "bicameral". After all, Jaynes himself is undoubtedly literate, a product of a literate society by whose action even the hypothetical connections in his brain are set up differently from those of illiterate societies and from which he developed his interpretation of a society without writing. The texts of a literate society (e.g., logical) cannot be translated into the language of any oral culture.

Rationality has its roots no later than Aristotle and his view of language (Harris, 2009). Gradually, language also came to be identified with a reduction to a written form that could then be transcribed, analyzed ("the number of strokes in a typescript") and interpreted in various ways, even identified with letters and syllables. And on top of this, there have been constant attempts (since the Renaissance) to introduce an artificial language with rational, clear and logical rules, without exceptions and metaphors, in the manner of a computer language (Eco, 1997). No wonder that the linguist Roman Jakobson (1971) compares DNA to language, much to the delight of many biologists who have since "known" what language is. Particularly after the introduction of the printing press, Harris says, we became a species of *Homo typographicus*. And Harris's fundamental question, which he addresses in many ways throughout his book, is: "Is it true, as has been claimed, that writing restructures consciousness?" Did the invention of writing either require a new mentality or have a profound effect on mental processes? Does the literate mind differ in important respects from its predecessor, the preliterate mind? Or is this a flattering deception fostered by literacy itself?" (Harris, 2009, p. 10). He then analyzes these questions from the perspective of "scriptism", i.e., the influence of writing on the conceptualization of speech; scriptism considers the basic properties of speech and writing to be identical, both being a set of established conventions translatable one to another. In a scriptural civilization, of course, records of all kinds play a central role, since they constitute an essential component of memory, both individual and social. Less attention is paid to the constant (re)interpretation of such records: they represent a complementary counterbalance. To summarize: the bicameral mind has been replaced by the rational mind; consciousness, philosophy, logic, calculus, and science, among others, have emerged.

Walter Ong also comments on the transition from an oral (speech sound-dependent) culture to a literate culture (focused on visual perception) in his now classics *Orality and literacy* (Ong, 2012). He too begins by referring to the Homeric epics, which allow us largely to reconstruct the tactics of the ancient bards and the demands on their memory. He notes that only a tiny fraction of the words in the *Iliad* and *Odyssey* are not part of standardized turns of phrase, and often such that are wonderfully predictable (Ong, 2012, p. 23). Moreover, these turns are associated with standard themes, which, along with the hexameter rhythm and chant, allows for repeated (though never quite identical) recitation (Ong, 2012, p. 59). Only later, through writing, does consciousness gain a distance that cannot be conveyed in any other way (Ong, 2012, p. 81). As literacy spreads, the perception of the world also changes, and this process culminated with the mass introduction of printing: "By removing words from the world of sound [...] and relegating them definitely to visual surface, and by otherwise exploiting visual space for the management of knowledge, print encouraged human beings to think of their own interior conscious and unconscious resources as more and more thing-like, impersonal and religiously neutral. Print encouraged the mind to sense that its possessions were held in some sort of inert mental space. Print encourages a sense of closure, a sense that what is found in a text has been finalized, has reached a state of completion." (Ong, 2012, p. 129) And elsewhere he points out: "The tremendous store of historical, psychological and other knowledge which can go into sophisticated narrative and characterization today could be accumulated only through the use of writing and print (and now electronics). But these technologies of the word do not merely store what we know. They style what we know in ways which made it quite inaccessible and indeed unthinkable in an oral culture." (Ong, 2012, p. 152) In the second half of the twentieth century, Ong believes that with the advent of new media (telephone, radio, television, film) there is another break in the thinking of our civilization, a secondary orality, with consequences that are hard to predict. These new media will also complete the notion of communication, originally an intersubjective process: written and media forms lack a communication partner (Ong, 2012, p. 173).

How all this relates to a possible transition to a kind of somatization, a shift of consciousness to a different mode of brain functioning, Ong does not and cannot say; the methods of his time did not allow it. Thus, John Hartley writes in the postscript to the 2012 edition of Ong's book, 30 years after the first edition (in Ong, 2012, p. 217), "His claim, that the historical shift from the technology of knowledge (speech) to another (writing) 'restructures' human consciousness, is a simple overgeneralization." But assuming that a change in thinking and perception of the world is reflected in a rebuilding of brain connections, such a change could theoretically be—at least as a correlation between biological and behavioral data. We select just one example where such testing might go; however, we note up front that this is a study of the activity of brain centers in individuals and not focused on heredity. Dehaene et al. (2010), using fNMR (functional nuclear magnetic resonance imaging), studied the activity of various brain centers in experimental subjects addressing a variety of psychological tests. The experimental subjects belonged to four groups: (1) illiterates, (2) literates leading the normal life of an educated person, (3) subjects who learned to read and write in childhood but led the life of a functionally illiterate person, and (4) subjects who learned to read and write only in adulthood. Large differences in the involvement of the brain centers (in the same tests) existed between group (1) and (2), with the other two showing a similar pattern of brain activity to group (2). Thus, it seems that mastery of reading and writing itself rewires the involvement of certain kind of brain centers and does so for life. We do not know, of course, whether people born in a culture that has known writing for centuries are hereditarily predisposed to respond actively to writing—as we pointed out above.

And one more opinion, quite pragmatic. Erich Fromm (2013) conceives of the history of the Jewish people as a process of mutual long-term cultivation of an originally nomadic people and their jealous God into a state where man is free and moral, and God is his partner and mentor. It's not as compelling a read as Jaynes, but just as joyful, and belongs in our narrative.

Let us conclude with some papers pointing out that the possible "set-up" of a literate culture may be collapsing in recent decades, with consequences that cannot be predicted.

Virtual Reality In the last few decades, *Homo typographicus* may be undergoing a transition to *Homo informaticus,* which is related to the introduction of computer technology, smart phones, the Internet, artificial intelligence, etc. In particular, child psychologists have warned that there may be a massive reorientation of children's and adolescents' perception of the world (e.g., Carr, 2010; Greenfield, 2015; Wolf, 2008, 2019). The leitmotif is a variation on Socrates's ancient concern about the introduction of writing, but this time it concerns not the introduction but the abandonment of reading: "Every time I watch my two sons complete their homework with the help of the internet, only to tell me later that they 'know all about it', a Socratic concern haunts me in relation to our culture" (Wolf, 2008, p. 77). Carr (2010) summarizes: children and young people struggle with reading (they are often functionally illiterate), they do not understand the text of, for example, a textbook, they are chronically distracted—and yet there is no turning back, we are entering a new and unpredictable period. The medium shapes what we see and how we see it, and quietly shapes individuals and society in a new way. And the medium has begun to transform itself in just two generations. Time will tell what this will bring, but what is certain is that the constant clicking to more and more links (even several dozen per hour) does not allow for focusing on a single text and understanding it, thus there is also no possibility of remembering it in its context. A fragmented hodgepodge of different messages floods into our minds, taking up space in short-term working memory without any of it having a chance to make its way into long–term memory. We become soaked in a multitude of different texts instead of keeping our attention on at least one. The role of hyperlinks—to attract our attention—simultaneously induce a continuous distraction from what we read. Again, according to many authors, this leads to changes in the connectivity of the brain centers or perhaps of both frontal hemispheres; more importantly, the process is global, it proceeds in *an* entire global civilization *at once*. A change in the medium of information supposedly also necessarily leads to a shift in the meanings of messages. We are not able to maintain deep knowledge that is inseparable from building once unique intelligence. For hours, often the greater part of our waking day, we (including young children) sit in front of a screen, resembling lab rats who constantly press a lever to get a pleasurable reward.

Let's leave aside the lamentations—we obviously lack the distance for a more thorough analysis. But the present is undoubtedly changing, and it is an open question whether civilization is currently charting a novel horizon of expectation (especially with the arrival of AI), perhaps accompanied even an arrival of a Lotmanian *explosion*? At such a moment, the perception of the world, and therefore the perception of the present as well as the points on the horizon of expectation, may be reconfigured.

What can we add? We could perhaps say that it is one thing to hear voices in the spirit, another to consider them real (coming from outside) and a third to feel them as imperative. From the point of view of historical semantics, then, the question is how the language of a given time allows new phenomena to be named (in terms of its vocabulary and metaphor). Jaynes discusses the transformation (of the way of experiencing) human innerness that was to take place—in terms of human evolution—in a historically short period of time (four millennia ago). Such a transformation would be consistent with what Barfield (1988) says about the historicity of phenomena. One can consider this transformation as having taken place in the sense of (c–e) above. And it can thus be argued that cultural development (exhibited by the changing nature, frequency and configuration of textual, artefactual and iconographic source traces) may (or may not) have played a role in it. In thinking hypothetically about the way in which this might have happened, we can draw on theories of symbolic (Cassirer) or semiotic representation (Peirce, Deely, Eco) with which we have repeatedly worked in previous chapters.

References

Ankersmit, F. R. (1994). Six theses on narrativist philosophy of history. In *History and tropology. The rise and fall of metaphor* (pp. 33–43). University of California Press.

Ankersmit, F. R. (2001). The linguistic turn: literary theory and historical theory. In *Historical representation* (pp. 29–74). Stanford University Press.

Assmann, J. (2002). Gedächtnis [Memory]. In S. Jordan (Ed.), *Lexikon Geschichtswissenschaft: Hundert Grundbegriffe* (pp. 97–101). Reclam.

Barfield, O. (1988). *[1957] Saving the appearances. A study in idolatry*. Wesleyan University Press.

Berlin, B., & Kay, P. (1969). *Basic color terms. Their universality and evolution*. University California Press.

Bultmann, R. (2019). *History and Eschatology: The Presence of Eternity*. Baylor University Press.

Carr, N. (2010). *The Shallows: What the Internet Is Doing to Our Brains*. Norton & Co.

Chartier, R. (1985). Text, Symbols, and Frenchness. *The Journal of Modern History, 57*, 682–695.

Chartier, R. (1996). *On the Edge of the Cliff: History, Language and Practices*. Johns Hopkins University Press.

Danto, A. C. (1965). *Analytical Philosophy of History*. Cambridge University Press.

Darnton, R. (1985). The Symbolic Element in History. *The Journal of Modern History, 58*, 218–234.

Darnton, R. (2009). *[1984] The Great Cat Massacre: And Other Episodes in French Cultural History*. Basic Books.

Deely, J. (2008). *Descartes & Poinsot. The crossroad of signs and ideas*. University of Scranton Press.

Deely, J. (2010). *Medieval philosophy redefined Scranton*. University of Scranton Press.
Dehaene, S., et al. (2010). How learning to read changes the cortical networks for vision and language. *Science, 330*, 1359–1364.
Durkheim, É. (2013). *[1894] The Rules of Sociological Method: and Selected Texts on Sociology and its Method*. Bloomsbury Academic.
Eco, U. (1997). *The search for the perfect language*. Blackwell.
Elias, N. (2000). *[1982] The civilizing process: sociogenetic and psychogenetic investigations*. Blackwell.
Fox, J., et al. (2017). The consequences of sibling formation on survival and reproductive success across different ecological contexts: a comparison of the historical Krummhörn and Quebec populations. *History Family, 22*, 364–423.
Fromm, E. (2013). *[1966] You Shall Be as Gods: A Radical Interpretation of the Old Testament and Its Tradition*. Open Road Media.
Geertz, C. (1993). *Interpretation of cultures*. Fontana Press.
Gellner, A. (2009). *Nations and Nationalism*. Cornell University Press.
Giddens, A. (1989). *Sociology*. Polity Press.
Ginzburg, C. (1989). [1980] Clues: Roots of an evidential paradigm. In *Clues, myths, and the historical method* (pp. 87–113). Johns Hopkins University Press.
Ginzburg, C. (2013). *The Night Battles: Witchcraft and Agrarian Cults in the Sixteenth and Seventeenth Centuries*. Johns Hopkins University Press.
Gleixner, U. (2005). *Pietismus und Bürgertum. Eine historische Anthropologie der Frömmigkeit [Pietism and burghers]*. Vandenhoek & Ruprecht.
Gould, S. J. (1996). *[1981] The mismeasure of man*. Norton & Co.
Greenfield, S. (2015). *Mind change: how digital technologies are leaving their mark on our brains*. Random House.
Harris, R. (2009). *Rationality and the literate mind*. Routlege.
Havlíček, J., et al. (2021). One but not two grandmothers increased child survival in poorer families in west Bohemian population, 1708–1834. *Behav Ecol, 32*, 1138–1150.
Hawkes, K. (2003). Grandmothers and the evolution of human longevity. *American Journal of Human Biology, 15*, 380–400.
Hayden, E. C. (2013). Taboo genetics. *Nature, 502*, 26–28.
Himl, P. (2010). Neosvícení a neposlušní. Regulace náboženské praxe a reakce na ni v Čechách ve druhé polovině 18. století (Českokrumlovsko Vimpersko, Strakonice, Žatec) [Not enlighted and disobedient]. *Dějiny—teorie—kritika, 6*, 163–221.
Iggers, G. G. (1997). Historisches Denken im 19. Jahrhundert. Überlegungen zu einer Synthese. In W. Küttler et al. (Eds.), *Geschichtsdiskurs, Band 3: Die Epoche der Historisierung* (pp. 459–470). Fischer.
Jakobson, R. (1971). *Selected writings II. Word and language*. Mouton.
Jaynes, J. (1990). *[1976] The origin of consciousness in the breakdown of the bicameral mind*. Mariner Books.
Kahneman, D. (2012). *Thinking, Fast and Slow*. Generic.
Kay, P., & Maffi, L. (2000). Color appearance and the emergence and evolution of basic color lexicon. *American Anthropologist, 10*, 743–760.
Kay, P., & McDaniel, C. (1978). The linguistic significance of the meaning of basic color terms. *Language, 54*, 610–646.
Kletvík, J. (2017). Evangelíci a spiritismus v Podkrkonoší. [Protestants and spiritism under Krkonoše Mountains]. *Z Českého ráje a Podkrkonoší, 30*, 62–107.
Koselleck, R. (2004). *[1979] Futures past: on the semantics of historical time*. Columbia University Press.
Lahoda, V., et al. (1998). *Dějiny českého výtvarného umění (IV/1) 1890/1938 [History of the Czech visual arts]*. Academia.
Lakoff, G., & Johnson, M. (1980). *Metaphors We Live By*. University of Chicago Press.

Lenneberg, E., & Roberts, J. M. (1956). The language of experience: A study in methodology. *International Journal of American Linguistics*, (*Suppl.* 22/2).
Lorenz, C. (2022). *Constructing the Past – An Introduction to the Philosophy of Chris Lorenz.* Princeton University Press.
Lottes, G. (2002). Neue Ideengeschichte. In J. Eibach & G. Lottes (Eds.), *Kompas der Geschichtswissenschaft* (pp. 261–269). Vandenhoeck & Ruprecht.
Mahlmann, T. (2004). *Hieronymus Zanchis Traktat* Über die Weissagung. *Narutwissenschaft die einem reformierten Theologen* [A treatise on divination by H. Zanchis]. In B. Mahlmann-Bader (Ed.), *Die Vermittelung alten und neuen Wissen in Literatur, Kunst und Musik, Band I* (pp. 315–348). Harrassowitz Verlag.
Maslowski, N., et al. (2015). *Kolektivní paměť: K teoretickým otázkám [Collective memory].* Karolinum.
Mayerová, F., & Vašíček, Z. (2004). Minulost a současnost, paměť a dějiny [The past and the present: memory and history]. *Soudobé dějiny, 4*, 9–32.
Medick, H. (1984). Misionare in Ruderboot? Ethnologische Erkenntnisweisen als Herausforderung an die Sozialgeschichte. *Geschichte und Gesellschaft, 10*, 295–319.
Meran, J. (1988). Historische Methode oder Methoden in der Historie? Eine Frage im Lichte der Methodologiegeschichte. In C. Meier & J. Rüsen (Eds.), *Historische Metode*. Deutscher Taschenbuch Verlag.
Mink, L. O. (1970). History and fiction as modes of comprehension. In *Historical understanding, Ithaca*. Cornell University Press.
Müller, H. P., & Sigmund, S. (Eds.). (2014). *Max Weber Handbuch. Leben – Werk – Wirkung*. J. B. Metzler.
Nešpor, Z. R. (2003). K původu a pramenům českého lidového spiritismu. *[Origins of popuar spiritism] Lidé města, 12*(2003), 59–81.
Nolte, P. (2002). Gleichzeitigkeit des Ungleichzeitigen. In S. Jordan (Ed.), *Lexikon Geschichtswissenschaft: Hundert Grundbegriffe* (pp. 134–137). Reclam.
Ong, W. J. (2012). *[1982] Orality and literacy*. Routlege.
Popper, K. R. (1988). *[1957] The poverty of historicism*. Routlege.
Repo, M. (2004). Sapiens dominabitur astris. Die theologiesche Astrologie Johann Arndts. In B. Mahlmann-Bader (Ed.), *Die Vermittelung alten und neuen Wissen in Literatur, Kunst und Musik, Band I* (pp. 349–368). Harrassowitz Verlag.
Richerson, P. J., & Boyd, R. (2006). *Not by genes alone: how culture transformed human evolution*. University of Chicago Press.
Ricoeur, P. (1984). *Time and narrative I*. University of Chicago Press.
Ricoeur, P. (1985). *Time and narrative II*. University of Chicago Press.
Ruse, M. (2017). *The Cambridge handbook of evolutionary ethics*. Cambridge University Press.
Simmel, G. (2004). *[1900] The Philosophy of Money*. Routledge.
Švorcová, J. (2023). Transgenerational Epigenetic Inheritance of Traumatic Experience in Mammals. *Genes, 14*, 120.
Švorcová, J., & Kleisner, K. (2018). Evolution by Meaning Attribution: Notes on Biosemiotic Interpretations of Extended Evolutionary Synthesis. *Biosemiotics, 11*, 231–244.
Třeštík, D. (2008). *Zápisník a jiné texty k dějinám [Notebook and other texts concerning history]*. Praha.
Troeltsch, E. (1931). *[1912] The social teaching of the Christian churches*. Allen & Westminster John Knox Press. Unwin.
Troeltsch, E. (2024). *[1924] Historicism and its problems*. Fortress Press.
van Dülmen, R. (1987). *Reformation als Revolution. Soziale Bewegung und religiöser Radikalismus in der deutschen Reformation [Reformation as revolution]*. Fischer Verlag.
Vašíček, Z. (1994). *L'archéologie, l'histoire, le passé: chapitres sur la présentation, l'épistémologie et l'ontologie du temps*. Kron.
Voland, E., & Beise, J. (2005). Bilanzen des Alters, oder: Was lehren uns ostfriesischen Kirchenbücher über die Evolution von Großmüttern? *Historical Social Research, 30*, 205–218.

Weber, M. (1972). *Wirtschaft und Gesellschaft. Grundriss der verstehenden Soziologie*. Mohr.
Wehr, G. (1995). *Esotherisches Christentum. Von der Antike bis zur Gegenwart*. Klett-Cotta.
Welskopp, T. (2008). Theorien in der Geschichtswissenschaft. In G. Budde et al. (Eds.), *Geschichte. Studien, Wissenschaft, Beruf* (pp. 138–157). Akademie Verlag.
West–Eberhard, M. J. (2003). *Developmental plasticity and evolution*. Oxford University Press.
Wittlich, P. (1982). *Česká secese [The Czech Art Nouveau]*. Odeon.
Wolf, M. (2008). *Proust and the squid. The story and science of the reading brain*. Icon.
Wolf, M. (2019). *Reader, Come Home: The Reading Brain in a Digital World*. Harper Paperbacks.
Zachar, O. (1914). *Z dějin spiritismu v minulosti [From history of spiritism]*. O. Zachar.
Žilka, F. (1941). *Naše křesťanství [Our Christianity]*. Laichter.

Chapter 7
General in History, or Slavery in Bohemia?

Abstract Discussion concerning the Czech early medieval society (tenth to twelfth century) will be presented on three casuistics, all showing how the ways of interpreting traces of the period may change the self–characterization of the society. (1) We shall discuss the ontological value of the "findings" of general comparative historiography: can we take analogical elements in several developmental lineages as of ontic nature, or do they represent a mere noetic effect? The question will be demonstrated by interpretative "introducing" of slavery into Czech history of that period. (2) We show how the Czech national community (via "its" historians) can restructure the space of its experience (i.e., the view of national history) according to its momentary needs: Whereas historians of the second half of nineteenth century interpreted the traces from Czech lands as an analogy to the situation in South Slavs, the generation of historians at the turn of the century coined the belief that since the tenth century the Czech history was already part of Western history. Thanks to such a shift of interpretation the image of the society considerably changed. (3) In the half of twentieth century, we witness yet another shift, when D. Třeštík interprets the traces per analogiam with the development in Poland.

Keywords General and comparative historiography · Slavic prehistory · Slavery in middle ages · Restructuration of the space of experience

> [Coleridge] fully grasped the part played by imagination in constructing not only the fictions of poets, but also the ordinary physical world which we speak of 'perceiving', though in fact we half perceive (that is, receive through sense-impressions) and half create it.
>
> Owen Barfield (1973), p. 27

A. Markoš, J. Horský, *The Seeds of Forms and the Forms of Seeds*, Biosemiotics 12, https://doi.org/10.1007/978-3-032-16131-4_7

7.1 Traces and Conjectures

In this chapter we will focus on selected issues of the early–medieval social and economic history of Czech lands. However, we will not be concerned with how things actually were at that time, i.e., in the tenth to twelfth centuries. Our aim is to show how the way of interpreting those conditions changed over the past two hundred years. The variation of interpretation shows how a community can read the same body of source texts in strikingly different ways under different cultural circumstances, in conditions of differently accumulated social memory and collective experience. By "community" we mean here "Czech national society" and "its" historians of the nineteenth and twentieth centuries (who wrote primarily in Czech, sometimes in German).

The number of source reports on Czech social and economic history from the early Middle Ages is rather limited. Interpreting them in such a way that a complete picture emerges can only be done, for example, by analogy with more familiar developments in other areas. However, the choice of area from which analogical interpretations are drawn is not purely scientific, it is a choice influenced in no small part by cultural identity. In the same way, the interpretation of fragmentary early medieval source traces requires to take at least an implicit position on questions concerning the generality of certain features of cultural and social development or, on the contrary, the diversity of its individual lines. Thus, we will see that around 1900, the question of whether slavery was common in the Czech early Middle Ages was also, implicitly, the question of whether Czech national society belonged (as a culture and as a civilization) to the Slavic East or to the West. In the same way, we see that for many historians around 1900, a rather strictly evolutionary idea that cultural development is almost a law of nature was in principle self–evident. We would like to consider several analogies between the evolution of life and the development of culture, or between biological and historical-scientific methods of interpretation. It turns out that a record, whether textual or genetic, *does not in itself have the capacity to determine (completely) how it is to be read.* It also turns out that in some cases it is not so easy to decide clearly between evolutionary necessity and historical contingency. A strict idea of the generality of development is sometimes determined by *default beliefs rather than a demonstration* (comparatively, inductively) of reality.

Just as Darwin's theory of evolution carried the notion of an acting force (natural selection) in the nineteenth century, some historians in the late nineth and early twentieth century relied on a distinctly directed tendency of general development—in the spirit of Spencer. Just as the strictness of Darwinism was toned down, cultural development was linked to cultural diffusion in anthropology, sociology and the cultural sciences thanks to Boas, and the comparative method came to be used more in a specific rather than in a general way thanks to Weber. Contemporary historians are thus much less inclined to general history than their predecessors of fifty or a hundred years ago. Therefore, before we get to the question of whether slavery was common in the Czech past, we will briefly address the debate about the possibility and legitimacy of general history.

7.2 Discussing General and Comparative History

The term *general history* is used in at least two senses. First, it refers to a specific *historical discipline*, and second, to the *actual continuity* and *branching* of the various lines of development within *human history*. The legitimacy of general history as a *scientific* discipline has been strongly questioned, if not rejected, in recent decades. This is often coupled with the misconception that if more general, multiple developmental tendencies are found when studying individual periods of developmental lines in human history, it would mean that history is governed by some *natural law*. Such a notion may seem incompatible with the paradigmatic and value-based foundation of the contemporary humanities. The existence of any such law of human development, even a supposed one, seems to contradict the conviction that only mutually equivalent cultural and historical forms of humanity can help us understand humanity itself. Moreover, such a supposedly recognizable law raises—and rightly, it must be admitted—concerns about its possible political and ideological implications. We believe, and in this chapter will try to show, that such an attitude towards general history, whether in the sense of the discipline or in the sense of the actual history of humankind, is often based on error and prejudice.

The error lies not only in the idea that we can do without general history. We will try to show that to some extent we are always grounded in it. The error lies in an unjustified fear of a kind of *naturalization* of humanistic, cultural and historical methods. Even then, as we try to show throughout this book "general history" may in some sense bring historical science closer to evolutionary biology.

Within biological, as well as some historical literature, *contingency* is sometimes attributed to history, but not evolution which is considered lawful. If we were to accept such a definition in a strictly logical form, it would at least implicitly seem that the defining feature of "evolution" is determinism. But this would not be a correct idea. For example, as Stephen Jay Gould (1989) points out, historical contingency is crucial to the evolution of life. Gould illustrates the role of contingency with the metaphor of a repeatedly replayed tape. Below and also in following chapters, we provide examples of research that attempted to induce a repeated or parallel replay in the laboratory, or identify one in nature—as a comparative study of individual evolutionary lineages exposed to similar conditions. The aim of these investigations is on the one hand to determine (in some cases experimentally) the strength of determinism, and on the other the degree to which historical contingency is present in the evolutionary processes of individual lineages. Biologists are thus thematizing problems and asking questions that have their parallel in historians' discussions concerning the legitimacy of "general history" and the meaningfulness of "comparative history". An example of this is the approach taken by one of the classic defenders of comparative methods in history, Miroslav Hroch.

Hroch strives for a "comparative approach" that is oriented according to *analogous historical situations*: "The assumption, of course, is to ascertain that the objects of comparison have undergone roughly the same development, even though for each of them the changes have taken place at a different pace and at a different time" (Hroch, 2016, p. 36). Hroch thus investigates national movements on which he applies such a comparative approach. If he aims at comparing identical (or partially different) courses of historical processes (i.e., a kind of chained configuration of phenomena) that occur at different times and have different temporalities in different places, then his research is analogous to studies in evolutionary biology studies that test Gould's metaphor of the replayed tape. Hroch's comparisons of analogous historical situations are related to the role that the categories of "necessity" and "contingency" (or even "causality") play in his thinking, and to the fact that he asserts that the goal of comparative studies is "to explain and understand historical processes" (Hroch, 2016, p. 10). The difference, presumably, is that studies following Gould seek to find contingency in evolution, whereas Hroch is looking for the general (i.e., presumably from his perspective, the "historically necessary") in diversity.

In Hroch's more recent studies (e.g., Hroch, 2016), a greater balance between generalizing and specifying applications of the comparative method is already evident. However, the attempt to *explain*, somehow and comparatively, the divergence of an individual historical line from a *general trend* is in actuality an attempt to exclude contingency. Here too an analogy with biological thinking can be drawn. Investigations within the framework of so-called evolutionary developmental biology (evodevo, Sect. 10.6) show that development (here in the sense of ontogenesis) is a key element in the evolution of multicellular life, influencing the relationships between genotype, phenotype and fitness. The role of individual ontogenetic development in evolution could perhaps, in historical lines, be represented by individual achievements. For example, we could consider the extent to which the individual life work of František Palacký[1] established the specific features of the Czech national political movement. In his work we find a specific, personal interpretation of the "cultural complex" called "nation" upon which the ontogeny of the Czech "national idea" is modelled.

Hroch's comparative method can be—and is—criticized in its generalizing role as a theoretical derivation from Marxist historicism. However, Hroch uses comparisons to explain the specificity of individual historical lines, reminiscent of Troeltsch's historicism or "moderated Marxism" (Troeltsch, 2024). The extent to which we accept or, on the contrary, quench criticisms of the generalizing use of the

[1] A nineteenth century historian, founder of the Czech historiography.

comparative method is closely related to the stance we take when addressing the question of the legitimacy of "general history", or the question of whether any form of it can be research–productive. These questions can be posed in both an ontological and a noetic sense.

Recently, Havelka (2019) attempted a consistent definition of the notion of "general history": Unlike Hroch (2016, 2019), he would reject the ontological anchoring of general history—for him, the ontological conception of general history would merge to a considerable extent with Hegelian or Marxist historicism. He concluded that Hegel "recognized as historical knowledge only what is common to events, what is actually necessary by its nature, and what is secured in principle" (Havelka, 2019, p. 267).

However, his critique of the ontological foundation of "general history" does not mean that Havelka rejects the existence of some common features occurring in multiple lines of historical development. "Alongside this, however, there are theoretically more explicit and methodologically more sophisticated 'shared histories', attuned to the parallels and structural analogies of certain civilizational, social and political processes, going beyond the primary differences of national histories, and especially 'comparative histories', with their intentions and conclusions also aiming at a certain generalizing reflection and going beyond a mere narrative of the antecedence, present and succession of history. Their operation would be virtually impossible without the systematization of certain general features present in the peoples and communities under study. The creation of so-called ideal types is still an important methodological means here" (Havelka, 2019, p. 269).

Let us say, however, despite the above criticisms, that "general history" in the sense of a scientific or historical discipline is not necessarily a "*universal* history", and that it does not even necessarily involve generalization. Let us consider "general history" as a discipline that seeks to define general concepts (mostly ideal types; Müller & Sigmund, 2014, pp. 63–66) by means of which certain invariants can be treated, taking the form of certain phenomena, structurations, or processualities that—in different and specific forms—occur or occurred in different places and at different times (Horský, 2019). In a sense, Georg von Below (1928) had already thought in a similar way about the appropriate sociological method of introducing concepts in historiography. The invariants under consideration can then take the form, for example, of Simmel's social "forms" (Kim, 2002, pp. 486–551) or the "functional equivalents" considered by Martin (Martin & Nitschke, 1986; Martin & Zoeppfel, 1989).

In his comparison—remaining based historical anthropology— Jochen Martin "softens" the Durkheimian category of "anthropological constants" with the notion of *functional equivalent*. Functional equivalents (such as the taboo of incest and distribution of gender roles, delimitations of the sacred, kinship systems, childhood, widowhood, etc., and even modes of distancing "us" from "them", and forms of domination and dependence of individuals and groups, etc.) are invariants that can be found in their particular variations in (almost) all known historical and contemporary cultures and civilizations. Individual cultures can then be compared according to such equivalents. The comparison *may* or may not reveal some identical or similar features or elements found in several cultures, even though it also reveals considerable cultural diversity.

If we want to look for parallels between thinking in biology and historiography, we could say that invariants of general history do not have their counterpart in sets of phenotypic (morphological and behavioral) features of individual species, but rather in phenomena on the level of, for example, body plans, identical within whole genera or higher taxa.

Contingency, as an example, may be another key concept common to historians and evolutionary biologists: let us therefore pay attention to it below (see also our glossary).

7.3 Analogous Phenomena in Multiple Lines of Historical Development

Whatever position we take on the question of the legitimacy and nature of "general history", it is hard to deny that in the history of human cultures, societies and civilizations, we can sometimes observe identical or at least *analogous* phenomena (synchronic or diachronic configurations), i.e., we can *find* cross-cultural analogies (Vašíček, 1994). We pronounce this with full awareness that this "observing" or "finding" cannot be left unproblematized. If we find any general features of historical development, of the processes of historical change, or of cultural transformation, this can be explained in at least three ways:

1. They can represent manifestations of truly general principles of development, which are immanently contained in the history of humanity and assert themselves despite all the evolutionary ramifications of human history in its individual civil, ethnic, cultural and national communities, etc.
2. However, the general unifying features of the historical process can also be interpreted in terms of the theory of *cultural diffusion*. Such "horizontal" transfers of a certain cultural variants can either be unintentional or intentional—and connected with missions, colonization, power, aggression, etc.

From the point of view of the attractionist conception of cultural development (as opposed to the selectionist one, including memetics), both cases above are always associated with an interpretative transformation of the transmitted cultural variant. At the same time, it is necessary to consider different levels at which cultural transmission can occur. The first involves transmission (only) on the *linguistic* (or conceptual) level: for example, legal and/or philosophical terms may come into use in another language without denoting the same contents, the same phenomena that it held in the environment from which they spread. Second, we can think of transmission in terms of (cultural) institutions, whether established by an explicit norm or merely emerging through cultural sedimentation. Examples of this include the structure of a large estate, monastic rules of this or that order, the technique of impressionist painting, or the style of Art Nouveau ornament. But this transfer is not necessarily connected with a genuine "accepting" of the cultural contents that had thus come from the donor culture. Finally, we consider the transfer of "cultural contents" (cultural variants and their complexes) that does have an impact on psychical structures—for example, the cognitive functions or behavioral manifestations of the members of the culture into which this transfer took place. Here, terminological or conceptual distinctions transferred from another culture (another language) do have an impact on differentiation within the field of sensations and perceptions, certain philosophical or educational texts can lead to psychologically formative and constitutive processes of the ability to reflect individual subjectivity or to the stabilization of habits, etc.

We can perhaps note that the distinction between levels of cultural transmission can contribute to critical debates concerning the linguistic paradigm in the cultural and historical sciences. Apparently, its proponents have postulated that only transfer at the level of language can be demonstrated. In contrast and each in a different way, Koselleck (2004) and the later Foucault (1990) both assume that words (concepts or discursive practices) are not only symptoms but also factors of sociocultural development. Sociopolitical language in Koselleck's terms or discursive practices in Foucault's terms are involved in the *modelling of the sociocultural field* or social practices.

3. But the uniform character of historical development, the appearance of similar (if not identical) elements in different lineages of cultural history can also be explained noetically as an effect of historiographic, historical–sociological, etc., *interpretation*. In explaining the course of history, we choose *explanatory models* that can be both *theoretical* and *narrative*. Thus, we work with general concepts (Weber's *ideal types*) or with more general narrative interpretations (and the *plots* integrated within them). At least in the case of some plots, the latter are always somewhat abstract. If we succeed in explaining the individual groups of cultural phenomena (sets of source traces) that belong to different lines of the historical development of culture by applying to them just one particular ideal type or just one particular plot, we thereby bring general features into the picture of these cultural contents—perhaps ontically independent of each other and more or less different in terms of the forms of specific lines of development.

According to Reinhart Koselleck, historians are always working in two directions: either they examine contents that have been previously articulated in the source language, or they reconstruct contents that have not been previously articulated in language, but which they pluck out of the remains by means of their hypotheses and methods. In the first case, they use concepts taken from the language of the sources as a heuristic input into the discussion of past events. In the second case, the historian uses concepts or scientific categories created and defined ex post—and used without being reportable in the source content (Koselleck, 2004). It must be noted, however: Koselleck is indeed saying that the historian "reconstructs" certain past "contents", where Weber (1904) would necessarily say only that the researcher "ideally-typically *constructs*", and where Ankersmit (1994) would probably say that the historian generates a certain interpretative context through the narrative interpretation. Both the ideal type and the plot, however, are always more or less abstract, and thus always *more general* than the content they describe.

This is also true of the concepts (or better—words, terms) that the historian takes from the source language. For example, the source term *servus* denotes in classical Latin (among other things) "slave". It thus refers to a certain "content", to a certain relationship of strong subordination. The term had spread in the early Middle Ages along with usage of Latin. It came to denote the status of various classes of dependent people for whom the national languages had their own designations, or for whom they did not yet even have a special designation. In addition to the term "servi", many medieval sources also use the term "mancipia". Old Czech used the term "rab", later also "otrok". These terms penetrated Latin written documents: for example, the term *otroce* from the eleventh century, which is interpreted by modern historians as a designation for the customs fee from the slave trade (Šusta, 1899). In relation to source texts from the nineth to twelfth centuries, Šusta (1899) says that "both meanings, rab and slave, express well the completely lawless position of the serf". This, however, is not a contemporary but a historical definition. Moreover, it is somewhat confusing, since at the time Šusta was writing, "serfdom" commonly referred to the position of bondservants in the Bohemian lands in the seventeenth to eighteenth centuries, as it lasted until the patent of 1781. To complicate things even further, last century's textbooks commonly referred the patent as having "abolished serfdom", although the word does not appear in that sense. Everything is yet further complicated by the fact that the historical term "second serfdom" was later coined to refer to serfdom in Central and Eastern Europe in the early modern period. In the Bohemian lands, the legal conditions for it were created as early as the sixteenth century, but in fact the situation that specifically Marxist historiography began to refer to as the "second serfdom" did not come into being until the use of serf labor on aristocratic estates in the second half of the seventeenth century (Maur, 2001).

Returning to the "serf" (or "rab" or "slave"), it can be thus said that in different places and at different times the term *servus* came to denote different forms of social interaction (different "contents"). Early medieval clerics, knowing Latin, compiling charters and writing chronicles, had the term *servus* at their disposal, which referred, to one degree or another, to a certain *abstract concept*. It is a question of historical semantics whether and to what extent a source author referring to people in a certain position as *servi* (for example, the inhabitants of a city, the urbs, would be the *servi* of the local bishop) changed the very *nature* ("content") of the social interactions involved or, on the contrary, to what extent the nature of the interactions so referred to, shifted the meaning of the word "servus", or increased the heterogeneity of the semantic field that is created around the term *servus*.

At the same time, however, and just as those of other eras, today's historians have some *pre-understanding* of the Latin term *servus*. This is due, for example, to the school education at the time or the way of reading special texts, but also to the more general "cultural climate" of the time in which the historian lives—including, as an example, the ways in which slavery is depicted in fiction or cinema. With such preconceptions they approach source texts written by Latin-speaking clerics hundreds of years ago. By confronting the source traces, they certainly correct their preconception, but always and at the same time—often unintentionally and implicitly—they introduce into the image of the past some traces of it.

Whether our interest in the term *servus* in the sources is purely philological, or in its potential as a legal term, or finally as a possible trace of former socio–cultural configurations, as scholars we will try to define a certain ideal type of "slavery" or formulate a certain plot scheme of "enslavement". Using these conceptual and narrative tools of interpretation, we will then attempt to bring some unity to the diversity of source accounts of "slavery". And this will be true both for those who explicitly mention *servi* and those who, although they do not mention *servi*, but—or so it seems to us thanks to our ideal types and plots of "slavery"—reflect correspondingly the conditions of "slavery". It is certainly true that, as Vašíček (1994) puts it, "historic notions" are "polythetic and diffuse" (i.e., overlapping and unbounded, which are inoffensive terms for vagueness). Therefore, their different meanings, different contexts and different uses—as well as their evolution—must all be taken into consideration. However, in order to use these in a controlled way, we must define their content in an ideal-typical way (Weber, 1904) and then detect deviations of their individual uses from this ideal type.

Let us then work here with the assumption that (not only human) capacity to experience the relation to oneself and to one's social and natural environment is closely related to—if not directly conditioned by—the extent to which we possess symbolic tools for experiencing and grasping them. The richness and variety of experienced (and therefore real) relations correlate with the subtlety of the differentiation of the symbolic systems of culture. This is, perhaps, most often exhibited in languages. However, it is necessary to consider also other sign systems carried by culture that do not have a linguistic nature. According to many writers—one of the first was Jacob Burckhardt—from the beginning of the Renaissance until the twentieth century there was an increase in differentiation within and between individual symbol systems in the West, and thus a potential thickening and finer differentiation of lived (i.e., socio–culturally actually real) psychic, social or environmental relations. Of course, relations outside the consciousness of the actors should also be assumed, but the question is whether any relations outside culture can be assumed after the "cultural turn" (see below). Especially since the 1990s, the idea that all relations—including those labelled as "social" or "economic"—are primarily cultural relations has also prevailed in the historiography of social and economic history. This is at the heart of the so-called "cultural turn" in the historical sciences (Štaif, 2006). Yet there remains the question of whether there are also "social", "economic" or "environmental" relations that are "outside" the system of culture shared by the actors under study. Weber (1991) says that what is "meaningless" is "nature". But as a whole, "social life" is broader than the domain of the "meaningful". This must, however, be understood in the context of Weber's noetic and sociological considerations. By "life" here is meant "empirical", i.e., historically occurring "reality" in its concreteness. The distinction *meaningful–meaningless* is conceivable only within a particular culture. So, Weber actually says that certain parts of "social life" can be seen as "nature", with both terms "life" and "nature" being used here in a different, narrower sense than is common in biology or biosemiotics.

We can therefore very likely assume that the actual experienced rational and emotional "content" of "slavery" as lived by early medieval people differed in different parts of Europe. It is just as appropriate to assume that the quality of "slave" as experienced by the actors of the time was different from the quality that modern Westerners conceptually and emotionally conceive of as "slavery".

Thus, we must always be cautious about whether we have discovered or, on the contrary, introduced any general features of the various lines of cultural history into images of the past. In the same way, a temporally former or spatially neighboring "same" can hardly be completely identical with the present and local "same". This is a reason for methodological caution, but not for resignation to general and comparative history.

7.4 Restructuring of the Ancient Past and the Self-consciousness of Modern Society

In one of his first historical studies, Josef Pekař writes, "Medieval economic and social life certainly is essentially the same throughout the nations of Central and Western Europe, and only by comparison will the meaning and essence of an institution stand out. Only from systematic comparison will it be possible to establish the interaction of nation upon nation and the distances in time by which nation differs from nation in its general development" (Pekař, 1901). This position deserves attention in several respects. We should focus first on (a) the "certainly essentially the same", then on (b) the formulation "throughout the nations of Central and Western Europe" and finally (c) Pekař's concern with "the meaning and essence of an institution."

Pekař, as can be seen, relied on the generality of basic developmental features. The question is whether, in his perspective, this generality (as specified in Sect. 7.3) is contained essentially in history of humanity, or rather caused by the spread of civilizational influences through cultural transmission. "Essentially" here would also mean in the sense of "evolution" understood as distinct in its causality from contingent "history". That Pekař may indeed have in mind an essentialist generality of development is suggested by the way he argues in his discussion of the type of rural community known as "zadruga", the term being a nineteenth century neologism used by Czech historiographers from the turn of nineteenth–twentieth century. These were cohabitations of up to several dozen adults that unfolded in a patrilineal and patrilocal manner. They had the structure of complex joint-family groups (Mitterauer, 1990) or were specific cases of multiple family households (Laslett & Wall, 1972). In contrast to Jireček's definition below, the main difference was that the "mayor" was not elected by the members of the zadruga, but the authority of the "householder" belonged to the oldest man in the community, and the authority of the "housewife" sometimes to the oldest woman thereof (i.e. not necessarily the wife of the householder (Budilová, 2019).

"The zadruga was also in Bohemia, because as a developmental stage it was everywhere: the 'Hauscommunion', similar to the present-day South Slavic zadruga, was undoubtedly common among the North-West Slavs in the first centuries of their historical (and also agricultural) life, but we have no work that sufficiently proves its existence" (Pekař, 1900). In the context of our considerations, Pekař's reliance on the general features of development in all the peoples of Western and Central Europe deserves more attention. This is a manifestation of a fundamental change in the interpretative strategy of Czech historians, a change that is probably related to the general "realignment" of the Czech socio–cultural field around 1900. Czech history of the nineteenth century also interpreted national history by incorporating it into a wider context, but this context was the Slavic world. Order was routinely brought to scarce source traces from the Czech early Middle Ages by interpreting them *per analogiam* with the more familiar conditions of other, most often Southern Slavs. This was done by working with the idea of some kind of Proto–Slavic (parallel to Proto–Germanic) family, clan or social relations, which were gradually disturbed by development—in Palacký's conception by "encounters and strugglings" with Germanic rules. Later Czech historians, who began publishing their first studies around 1900, came up with a fundamentally different interpretative scheme. They interpreted history of the Czech state, society, economy and law, at least since the tenth century, through analogy with the development of the basic institutions of the Latin–Christian world. If historiography contributes in any way to the formation of the self–confidence of the (national) society from whose soil it grows, their position might be as follows: Since the beginning of our national history in what is now Czech territory, we have been part of the West. Moreover, the explanatory scheme of the gradual erosion and loosening of Old Slavic archetypes is replaced by an explanatory scheme working with the idea of (general) social and cultural developmental stages. Modern (national) society, in the thinking of its historians, no longer seeks basic values and existential security in an idealized antiquity, but in a forward movement and in participation in general human civilizational and cultural "progress". *Essentially the same corpus of source texts became read in a new way, a new interpretative arrangement is introduced into more or less the same sum of source traces.*[2]

An instructive example can be the development of the discussion on the "Old Czech zadruga". One of the classics of nineteenth–century Czech legal history, Hermenegilt Jireček (1863), says that "a zadruga is a union of persons who, being united by kinship, have a common farm and live under an elected mayor, working together and sharing food resources" (Pekař, 1900, p. 253). Very similar-looking engagements were described by Slavic ethnography especially in Serbia throughout the nineteenth century. The idea that what can be observed ethnographically in some

[2] A biosemiotic paraphrase would sound like this: *Essentially the same corpus of genes became read in a new way, a new interpretative arrangement is introduced into more or less the same sets of genes.*

South Slavic areas is a remnant of the primordial Slavic conditions has gained ground. If this is the case, then it is possible, even necessary, to use the arrangement of the zadruga as a means of interpreting fragmentary reports from the early medieval history of other Slavic societies, including that of Bohemia. Jan Erazim Wocel argues, convincingly at first glance: "If we consider that visible traces of this patrilineal and hereditary order, based on the equality of the individual members of the family, are found in all Slavic peoples, we can thus see that such an order and manner could not have been formed in the historical times of the Slavic peoples, scattered into various regions, but that its origin goes back to those ages obscured by the pre-historical fog, where the Slavs led a common life in their original homeland and cradle. For it certainly could not have happened that in some historical period the Russians, Poles, Czechs and South Slavs together—perhaps at some assembly—agreed on the establishment of such an identical order and, exchanging their views, finally established the law which, as far back as the oldest sources go, has prevailed among all Slavic peoples in lands far apart, and has prevailed in part to this day" (Wocel, 1861, p. 7).

The persuasiveness of Wocel's interpretation is, however, due to the fact that these "traces are not visible" in themselves, but only after their interpretation by analogy with the then recent zadrugas. Villages were to emerge gradually from individual zadrugas, as some of the existing members withdrew from them. The original common ownership and common cultivation gradually became collective ownership, but with temporary individual cultivation. Subsequently, the temporary possession of the land by individual families turned into permanent ownership (or later into serfdom). The more recent medieval sources (from the twelfth–thirteenth centuries and younger) were as if confirming traces of this transformation process. For example, the serf villages in the original settlement areas (as opposed to those that were established only after medieval colonization) had communal land (*communia*) of pastures or meadows. This communal land was seen as the last relic of the original common ownership of all land by the zadruga. Another symptom and remnant of the zadruga was thought to be a specific right of inheritance, i.e., a "(family) undivided share" or "common" (Kadlec, 1898). From the 1870s onwards, the zadruga theory could be linked to the belief in the generality of the so-called agrarian communism among the Indo-Europeans in their early agricultural days.

In the 1890s, there was a significant change in the interpretation of the earliest agrarian history. Jan Peisker (1899) made a groundbreaking claim that the South Slavic zadruga came into being only in the Middle Ages under the influence of the Ottoman Empire's tax system. The fact that the Yugoslav zadrugas are of a relatively recent date has likewise been acknowledged in contemporary literature (Budilová, 2019; Todorova, 2001). Peisker places the origin of the Yugoslav zadrugas in an even more recent period than Pekař does. But what is significant for us at this point is the distinctiveness of the analogies of the interpretative principle applied by Peisker and Pekař. Pekař finds arguments in favor of assuming a former existence of the "Hauscommunion" in Bohemia "in the analogy with the *German* development" (Pekař, 1900, p. 265). He expresses the zadruga as a concrete form of a more generally widespread developmental stage. This argumentation was reinforced at the time, for example, by Dopsch's (1909) interpretation that the zadruga—or similar large, jointly held farming family groups—emerge in areas with difficult agricultural conditions (for example, among Alpine Slavs on peasant farms at higher altitudes). In addition to rejecting the claim of an archaic nature of the zadruga, Peisker (1899), also rejects the theory of primitive agrarian communism among the Indo-Europeans. He argues for agrarian practices from the early days of agriculture. He concludes from ethnographic research and from written sources, which are not entirely clear-cut, that the right of free land use existed at the beginning of agricultural life among the various Indo-European tribes. According to him, there was land enough everywhere. Anyone could clear and carve out a piece of land. So long as the land thus improved was productive, people were free to use it. After they had left it, someone else could take it. According to Peisker, the agrarian conditions of this so-called cyclical agriculture make questions of whether ownership of land was primarily individual or collective meaningless.

Pekař's concern with the "meaning and essence of the institution" still deserves more detailed commentary. The research path that takes historians to accept such meanings and essences led and leads—with reference to Koselleck's distinction quoted above—both through concepts taken from source language and through legal, anthropological, sociological, etc. categories of extra-source origin. There was and still is a debate as to whether the nature of these institutions is inferred through generalization from particular cases (corresponding to Simmel's "formal sociology"), or whether it is conceptually constructed in a "pure" form and particular cases are measured by this construct (in the sense of Weber's "ideal types").

Nineteenth-century authors tended to define "institutions" primarily in legal terms. For example, the nineteenth century studies that addressed the question of the occurrence of zadruga in the Czech early Middle Ages were primarily historical-legal studies (Kadlec, 1898; Wocel, 1861). A specific form of property and inheritance law, the so-called "(family) undivided share", was not only considered a possible symptom of zadruga, but was directly identified with it. In contrast to these earlier works, Peisker (1899) also sees an agrarian dimension to the institution of zadruga, while Pekař (1901) sees its "social" or "economic" dimension. Such an approach corresponds to the then newly established sociological approach. For example, Weber, in a study from the early twentieth century, distinguishes between the "juristic" and "empirical" validity, the scope of a particular legal concept or legal order. Sociologically considered, law co-determines social agencies and relations insofar as it enters an "empirical" (i.e., in a given situation, actually intended by the actor) "meaning of action" (Weber, 1991, pp. 160–162). Thus, a sense of the cultural dimension of institutions, including their semiotic dimension, gradually asserted itself.

One of Josef Šusta's earliest scientific studies, *Slavery and large estates in Bohemia*, from Šusta, 1899, fits into these transformations not only as a symptom, but above all as an important factor in them.

7.5 Slavery in the Czech Early Middle Ages

Historical development can appear to us—as mentioned above—as a general, uniform development along individual lines, either because it really *is*, or because it gradually *becomes* so through cultural diffusion, or finally because we *interpret* it as uniform. The question on the social ontic level is what actually happens in terms of meaning when some specific cultural (legally defined) institution is transferred into a field in which active or activatable relations differ from those in the field from which the institution comes. There is certainly a tension between certain abstractness of such a cultural (legal) institution or this (legal) concept and the concreteness of both the field in which it originally operated and the field into which it is transferred. Thus, the researcher's notion may mistakenly assume qualities (contents) that were not present in the time under study— "yet" in terms of development, "never" in terms of cultural diversity. In working terms, we can therefore distinguish at least three levels of consideration of a given question. We have here its *historical-semantic dimension*: the word-terms "servus", "rab", "slave" may have changed their meanings. We have the *historical–semiotic* or *socio–cultural* formative dimension: the term, or the institution it designates, has a different formative capacity in different socio–cultural fields. Finally, at the noetic level, researchers address whether and to what extent the observed phenomena are in principle *identical*, or whether interpretative clarity can be brought to the phenomenal diversity through interpreting fragmentary source traces in reliance on general developmental tendencies.

With all this in mind, let us turn to Šusta, 1899 study *Slavery and the large estates in Bohemia*. This study is groundbreaking, although Šusta was not the first to consider the abundance of slavery in early medieval Bohemian history. Šusta's article, however, has "slavery" as its central theme. The very first paragraph deserves attention:

"In times of advanced education, slavery is an unpleasant social anachronism. In times of primitive culture, however, it tends to be a necessary institution and an important means of economic progress. The very fact that a conquered enemy is not killed but exploited economically is a progress. A flourishing of agriculture has always been connected with slavery among all peoples renouncing hunting and nomadism. The warlike hunter and herdsman never took the scratch plough himself; the work was for a long time left to women and slaves alone. Slavery, too, had its cultural mission, and disappeared only after it had been fulfilled" (Šusta, 1899, p. 34).

Šusta's conception is purely *causal-evolutionary*, relying on the *lawful generality of development*. In the context of the transition to agriculture, slavery was "among all peoples", it fulfilled its *cultural mission*, only to be subsequently historically superimposed.

In the very next paragraph, Šusta draws implications for Czech history: "The period of flowering slavery constitutes a completely regular phase of the social development of individual nations, and therefore there is no need to conceal or diminish its existence in national history. It has been unjustly done more often in

Bohemia. As the occurrence of slaves in our old records could not be entirely denied, the testimony is at least weakened by the assertion that slavery in Bohemia was never quite the settled condition of whole classes. But even this statement is not correct" (Šusta, 1899, pp. 34–35).

According to Šusta, the weakening of the significance of source reports that can be interpreted as evidence of the occurrence of slavery was based on some Byzantine sources. However, Šusta considers this to be an incorrect approach. He says this in the paragraph quoted below, which is a stark contrast to Wocel's position above on how to arrive at a correct picture of the earliest Czech social history:

> Social conditions are everywhere formed by the country inhabited and its nature, and not by linguistic and tribal affinities. Economic development has always taken a different path in the wide plains of Hungary and South Russia than in the forested highlands of Central Europe, even though in both these regions settled peoples related by blood and language. The affinity of language has always had only a very slight influence on the development of culture. Therefore, it is not correct to extend the judgements made about one Slavic tribe immediately to other tribes living in a different environment (Šusta, 1899, p. 35).

Šusta wrote these words at a time when cultural anthropology was giving birth to its linguistic and cultural relativist trend, which began to identify language to a considerable extent with the symbolic universe of the whole culture. All the more interesting is Šusta's assertion that "the *affinity of language* has always had only a *very slight influence* on the *development of culture*". Apparently, for Šusta, "language" and "culture" did not coincide, or at least not to the same extent as for the representatives of linguistic relativism. Furthermore, geographical conditions are, according to Šusta, a strong factor in historical development. Social conditions are formed by the nature of the land a given society inhabits. To make it completely clear that he is changing the basic interpretative paradigm compared to the classics of nineteenth century Slavic history, Šusta adds: *not linguistic and tribal affinities*. According to Šusta, both have a clear implication: the gaps in the source reports from the Czech lands cannot be interpreted *per analogiam* to the general (South) Slavic situation. He writes:

"The economic conditions of the Czechoslovak countries are much closer to Western Europe than to the Slavic East. The course of Czech cultural history is attached to the Western world. That is why it is impossible to use South Slavic records as sources for Czech history. Especially in the questions of social history it is necessary to limit oneself to the sources that arose in the Czech lands" (Šusta, 1899, p. 35).

This is certainly the formulation of a basically innovative methodological starting point. But it is also, at least implicitly, a *non-scientific cultural–value judgment* that establishes the direction of research, as Weber would say, or of determining the meaning of a historical event, or even of rhetoric, as Ginzburg (2013) would add. We could consider this a manifestation of the aforementioned self-awareness of modern Czech culture, which does not seek its support in an idealized (supposedly "democratic") Slavic antiquity, but in its inclusion in the general, progressive development of the West. Let us just note that the need to contrast the general developmental tendency with the assumption of an Old Slavic idyll is also encountered in

Jan Slavík's statement: "The medieval division of the Czech nation into nobility and subjects seemed to many scholars a retrograde step against the so-called democratic Old Slavic orders." However, Slavík says with determination that "democracy, where there were supposedly no social differences, did not exist among the ancient Slavs" (Slavík, 1940). Therefore, "we need to get rid of the misconceptions that arose mainly at a time when the lower classes, especially the peasantry, were fighting for liberation from serfdom. At that time—in the eighteenth and nineteenth centuries—there was a widespread opinion that the nobility had grown up as a harmful parasite on the body of the nation, which it had "deprived of freedom and eaten away'" (Slavík, 1940). In contrast to the idea of a Slavic archetypal social order, Slavík puts forth a general historical perspective:

"Under such conditions of life and society, the division of the nation into an innumerable ruling class and a mass of subject people was a work of natural development. This is generally evidenced by the fact that the same process of social division occurred in all medieval states, including, of course, all Slavs" (Slavík, 1940, p. 40).

But let's return to the article by Šusta: he finds slavery in Bohemia mainly in the ninth to twelfth centuries. In the ninth century, slaves were rather exported from Bohemia. In the tenth to twelfth centuries, with the development of agriculture, they became a common and abundant part of society. This is based on a report from Bavarian marketplaces, where customs fees were set in the ninth century on the sale of slaves (de servo) or women-slaves (de ancilla) imported from Bohemia (Šusta, 1899, pp. 38–39). The export of slaves from Bohemia was the first:

> For in a nation that does not yet practice intensive ploughing, slavery cannot be practiced on a large scale. The condition for a solid slavery is that the slave should support himself by his work and still give the surplus to the master. In a hunting nation, on the other hand, the master would have to support the defenseless slave. Hence the greatest part of the male captives was exported abroad, and thus their price fell. On the other hand, the female slave found a permanent place in the harems of the great noblemen much more easily, and her knowledge of textile work was highly valued among primitive peoples (Šusta, 1899, p. 39).

A change in the Bohemian lands occurred in the tenth century with the introduction of the system of the large estate. This created also a place for slaves in domestic economic conditions. "All the land of the large estate was divided into parts: 'salland', a part directly cultivated by slaves on the master's account, and 'hufenland', a part distributed to free subjects for the use of a fee. The center of the estate is the central court, where the goods in kind are collected, and where the multitude of slaves do all the industrial work for domestic use and sometimes for the market. Thus, each estate forms a closed economic unit for itself" (Šusta, 1899, p. 41). Šusta then states: "The period from the end of the tenth to the end of the twelfth century is the heyday of slavery in our country. Slavery appears in it as a regular essential part in the economic structure of a large estate; slaves are not merely a source for export as in the past, their labor power finds full application. Therefore, they become an object of import" (Šusta, 1899, pp. 39–40).

While the opening passages of Šusta's account sound, as mentioned above, almost causally evolutionist (for example, he generally links slavery to the beginnings of agriculture), when he describes the link between slavery and the landlord system, he moves more towards a diffusionist conception. (It should be noted that at the time when Šusta wrote his essay, the diffusionist theory of cultural development had not yet been fully formulated.) He says:

> The position of the nobility and clergy of the whole of Western Europe was economically based on a large estate of this kind, and such a large estate could be found in Bohemia as early as the time of the older Premyslid dynasty. The Bohemian princes were forced to adjust their economic conditions in the same way as they had seen in their German neighbors. Their wives and the clerical orders introduced from Germany were the most important factors in this reception. Thus, it came about that the same forms as in Western Europe prevailed in Bohemia on the clerical and ducal estates (Šusta, 1899, pp. 41–42).

Another passage also reverberates with the sense of cultural (Christianizing and civilizing) transfer: "While the medieval view of selling captives into slavery was not in principle opposed, Christians were not to be sold. But whoever stood outside the protective circle of the Church was an enemy and a lawless subject. In this view, the course of the whole cultural development of Europe in general is shown. With the Roman empire and Christianity, the circle of legal and cultural concentricity expanded from the Mediterranean onwards, and the barbarians lying outside of it, and therefore the lawless became fewer and fewer. It was not until the last century that the circle embraced, by the discovery of the natural rights of man, all mankind, not excluding the Turk and the black slave in the plantations. In the Middle Ages Christianity was that circle; whoever was outside it was without the natural rights of man" (Šusta, 1899).

The early medieval landlord system was closely related to subsistence farming: "In the 13th century the period of purely natural economy in Bohemia came to an end" (Šusta, 1899, p. 97). Most of the serfs' obligations changed from natural to monetary. At the same time, according to Šusta, slave labor ceased to be used, and people in the position of slaves disappeared from the source records.

So, what source reports did Šusta consider to be evidence of slavery? He had at his disposal essentially the same reports from documents, chronicles and legends that we have today. First, the above-mentioned reports on the sale of slaves or the donation of captives (presumably to slavery)—for example, he notes that the monastery of Břevnov had "a registered right to 'decimum hominem captivum' from the duke's plunder" (Šusta, 1899). On the other hand, there are occasional reports that some freemen (landowners rather than free peasants) had 'slaves' on their estates as laborers or as tillers of the soil. For example, we learn from documents that one of the free "heirs" (heredes, or milites secundi ordinis) was, after the amnesty of the throat punishment, "with his six slaves, given in servitude to the Břevnov monastery" (Šusta, 1899). In 1115, for example, an "heir" is reported, who "has four ploughlands in different places, which he could not have cultivated himself without slaves", as Šusta judges. However, we also know of freemen who, for example, as of 1143 *voluntarie se servitio subdiderunt* (Šusta, 1899). Finally, and in a way above all, many eleventh and twelfth century documents confirm the donation of (probably serf) villages or parts of villages, in addition to the donation of specific people (sometimes with wives). These people are either listed by their personal names or by their trade (their profession or rather specialization): "All these craftsmen were serfs, and especially this detailed division of labor is a testimony to the prevalence of slavery in Bohemia in the eleventh and twelfth centuries" (Šusta, 1899).

But how are we to imagine the true nature of the status of slaves in the early Middle Ages? The source traces are often in the nature of linguistically or legally configured terms, but these are, after the passage of eight or ten centuries, largely open—if not empty—in meaning. *Today* we can only interpret their content with a considerable degree of uncertainty. If someone was sold or given as a slave (*servus*) or slave girl (*ancilla*), we can assume that he or she was probably closely and strongly attached to the household or farm of his or her master. But if someone was a "slave" on someone else's "landed estate" (*terra ad aratum*), weren't they in fact just serfs, or serfs along with their families? And what might have been the position of the "blacksmith", attributed, according to Šusta's interpretation, as a "slave" to the court of some large estate? If someone voluntarily "servitio subdiderat", we can certainly translate this to mean that they voluntarily gave themselves in slavery. However, it can also be translated that they "voluntarily submitted" or "voluntarily undertook to serve". If they were free peasants who thus took on a subordinate position with their lands, what was the actual change about their position? This change must probably be considered in layers, at least in terms of legal status, social status and actual management. We should be cautious and avoid bringing into the ninth to twelfth centuries "contents" (qualities) that only reveal themselves to the modern spiritual world of the liberal individualistic West.

Just as the sources give us no certainty as to the meaning of words such as *servus*, neither do they provide us with any detailed account of the actual workings of the "large estate". In the context of Šusta's interpretation, we can consider the double dimension in which the large estate is introduced into Czech history. First, as Šusta assumes, in the sense of cultural transfer by the actors, and second in the sense of scholarly interpretation. On the first level, the above-mentioned question remains open as to what extent this was a mere naming of the domestic situation with new Latin terms and to what extent this new naming was also a factual restructuring of the socio–cultural field. This question can hardly be answered with certainty. On the second level, it is worth pointing out that "large estate" and "bringing into slavery" are not so much Šusta's conclusions, but rather tools of interpretation. In Weber's (1904) sense, the structured system of the large estate is an ideal type; in Ankersmit's (1994) sense, enslavement is a plot integral to narrative interpretation. They are a priori tools of interpretation. Without them, Šusta would not have been able to obtain such a clearly and distinctly formed picture of "slavery in Bohemia".

7.6 The "State" as a "Wise Organizer"?

But were all those whom Šusta regards as slaves really in slavery? What if we were to apply a different interpretative scheme to the same set of source traces? As indeed we can. Just such a scheme was offered by Krzemienska and Třeštík (1964) when they applied the structure of the Polish princely "servant organization" (or network of "servant settlements"), described in detail by Grodecki as early as 1913, to the Czech source material as an interpretative tool. This allows them to consider many of those who Šusta (but also others) would describe as slaves in an analogous sense of the word—as unfree—but nevertheless not as closely tied to the prince (or to the ecclesiastical institution to which the prince gave them) as slaves in the true sense of the word. These are at least some of the people who the charters list by their occupation (profession, specialization) as "servants" (ministeriales) who perform their trade as a servile service (ministerium, oficium). These ministeriales were likely originally in a position close to slave-like, tied to individual courts on large estates. But over time, according to Krzemienska and Třeštík, the prince began to settle them in certain villages so that, in the conditions of a subsistence economy, their service and labor would cover his economic needs.

It is "difficult to get an idea of what the social and legal status of [these] servants was like" (Krzemienska & Třeštík, 1964). From the documents we have reasons to believe that the prince could have given them away at any time and that in the process of giving them away he could have divided their families (given their wives or daughters to someone other).

"The prince gave them away on all sides, with land and sons, and he gave their wives with daughters and immature sons especially for service in the gynecees. It would seem, therefore, that their position was close to that of slaves, but the fact that they sat on the land and that, finally, at least some of them must be assumed to have special valuable expertise and skills must have guaranteed them a better position, approaching that of vassals" (Krzemienska & Třeštík, 1964).

"With relative certainty" Krzemienska and Třeštík (1964) say of those who "for the most part they were unfree, they were 'servi' as opposed to 'liberi'". However, "from this unfreedom they could redeem themselves and put another servant in their place, who was then of course a 'servus'. Although unfree, at least some 'servants could therefore accumulate considerable wealth'. By 1078 we have documented a ransom from such a position of 300 denarii, i.e. about one grzywna of pure silver" (Krzemienska & Třeštík, 1964).

What, then, was the society of the Czech state in the eleventh and twelfth centuries as conceived by Krzemienska and Třeštík? There were freemen, vassals and slaves, but there were also those specific servants. There were "large estates" of both princely and ecclesiastical institutions and probably also of the noblemen, but there was also an organization specific of peasants, i.e., princely "servant settlements". What arguments support this picture? Krzemienska and Třeštík begin their study by quoting a passage from the Chronicle of Cosmas, in which "Libuše [...] warns the people, who unwisely ask for a prince, against his rule", against what the prince will restrict them in and who he will seize. They then say:

> Cosmas borrowed this speech from the Bible, where Samuel similarly warns the Israelites against the rule of the king. He has, however, modified the list of services and servants to suit the Bohemian conditions of his time. Servants of the same occupations are found in contemporary and a few later documents. [...] We find here millers, bakers, shoemakers, furriers, fishermen, shepherds of all kinds, etc. But Millers, Smiths, Fishermen and

> Fishmongers, Sheepdogs, Sheepmen, Vineyards, etc., also occur very frequently as place-names. This group has long since aroused the interest of historians seeking to use toponomastic material as a historical source, but it has taken longer to recognize the obvious connection of these names with the names of the 'oficii' in the documents, however striking and at first sight conclusive the connection may be. Another rather striking fact is that this toponomastic type is not limited to Bohemia, but is very frequently encountered in Poland, Slovakia, the southern Slavs and Hungary. [...] All these connections are undoubted, but opinions on their interpretation differ considerably" (Krzemienska & Třeštík, 1964).

Krzemienska and Třeštík's interpretation combines traditional historical analysis of documents and toponomastic analysis with historical-geographical analysis. The connecting of the source traces from the Czech territory is then facilitated by analogy with the Polish situation. At the same time, the "theory of the state" is present in their interpretation (mostly implicitly), attributing an important economic role to the state. According to them, Grodecki, whom they are inspired by, "saw the cause of the emergence and existence of the service organization in the economic needs of the early medieval state" (Krzemienska & Třeštík, 1964).

In fact, when we input into the map the toponomastically interesting names of servant settlements, data on (donated) servants from deeds, princely fortresses, castles and courts (Krzemienska & Třeštík, 1964), we find "that settlements with servant names occur grouped in certain units. In these units, servants are also directly attested in documents along with their servant names. They can be found in villages scattered throughout the country, but if we summarize them all, we see that they are grouped in units together with the servant names." This can be considered as "proof" that "servants were indeed proper to the princely household and that church institutes received servants only from the prince" (Krzemienska & Třeštík, 1964). Further, they state:

> The princely service of wine-growers, fishermen and beekeepers is represented in the sources in great numbers. Winegrowers appear in the sources with vineyards and land, but mostly as an accessory to a donated vineyard. Along with them, other specialists necessary to service the vineyard, such as coopers and grafters, are also donated. References in the sources are also illustrated by numerous toponomastic names that spread throughout the old settlement area.

This leads them to the conclusion that the Bohemian prince (generally a landowner) "settled servants and craftsmen on the land around the courts and forts and collected from them part of their products or demanded part of their labor for performing various services and production tasks at the courts or at the forts. This is how the specialized villages that are reflected in toponomastics came into being" (Krzemienska & Třeštík, 1964).

Dušan Třeštík repeatedly returned to the theme of servant settlements. For example, the work Merhautová and Třeštík (1984) introduces the interpretation of the history of Romanesque art in the Czech lands with a description of the establishment of the Přemyslid state and its organization. It explains how the Bohemian prince gradually gained control over the territory of the Bohemian state and how he subjugated its population and secured tax revenues. Třeštík may be amplifying his interpretation here: "Taxes and levies, however, could be far from sufficient if their collection was not sufficiently effective. Therefore, the Přemyslid rulers were forced to organize their state perfectly, especially economically. The state had a given and essentially finite number of resources at its disposal and had to manage them in such a way as to satisfy all the needs of a predefined or estimated consumption. They therefore had to 'plan' in a highly complex way" (Merhautová & Třeštík, 1984).

The ruler established his provincial castles and extorted from (still or formerly) free peasants. At the same time, he organized crafts and services at castles, forts and courts in the conditions of a still largely subsistence economy.

> The prince also solved this problem simply and effectively: he settled servants and craftsmen on the land around castles and farmyards and collected from them a part of their products or the required part of their labor for the performance of various services and servitudes and imposed on them 'in perpetuity' a special 'service'. Specialised villages of one occupation were created, which were sometimes named after their specialisation. Toponyms all arose in the early Middle Ages as a product of this special, so-called service organisation (Merhautová & Třeštík, 1984).

Let us note that this interpretation sounds even a bit more explicit than the previous original from 20 years. *All* the settlements with servant names are said to have originated in the building of the servant organization in the tenth to twelfth centuries. The nascent state is said to have been *perfectly organized and planned in a highly complex way*. Reading these claims and comparing them with the disputed and—above all—very brief source traces, we cannot help but recall Koselleck (2004) warning that sources alone can never tell us anything about the past. We always need some "theory of possible history" that will allow the sources to speak. The idea of a state whose ruler intends to create a service system in an organized and planned manner, and to name the villages, so as not to confuse them, according to the specialists he places in them, is *in no way derived from the sources*. This idea is an extra-temporal concept built into the (partly implicit) theory of possible history. The relevant set of source traces can only say either "yes" or "no".

Box

Sources for biologists include, for example, sediment layers with fossils (trilobites were earlier than dinosaurs). Proxies also come into play—as in geology and planetology—and act as sources. Examples include bubbles of fossil air in a glacier, the occurrence of trace isotopes in preserved teeth or rocks, fossils and their sequence in geological layers, chemical anomalies in rocks indicative of life activity, etc. Network of such proxies, often confronted with present-day conditions (conservative biochemical, ontogenetic, ethological processes. etc.), allows inferences and analogies to build up an idea of things past. Let us recall yet another network of indirect data, let us say "sources preserved in palimpsests" which can be revealed, for example, by a comparative analysis of the sequences of recent records in biopolymers, in the body structures, in metabolism or in ontogenetic processes. Such an analysis can, for example, establish that the giraffe evolved from short-necked animals rather than being their ancestor. They often allow the reconstruction of branching lineages, sometimes even estimating their localization on a time scale. Similarly for translocations, hybridization events, etc. It makes it possible today, for example, to discover "what all" we inherited through our ancestors interbreeding with Neanderthals. Palimpsests of a similar kind, as we have seen, are also widely used by historians and historical linguistics.

But do the source traces really vindicate the theory of servant organization? More recent work, for example that of Jan Klápště, is more skeptical about this question. Klápště points out that "the intersection of the two sets, the written evidence of 'specialists' and the corresponding place names, has remained strikingly small" (Klápště, 2012). On the one hand, he does speak of an ingenious theory of servant organization, and he apparently acknowledges that in some cases the source reports can be read in terms of Třeštík's interpretation. "Yet, the theory of the servant organization is certainly not overloaded with arguments, and next to its light construction, every factual question seems partial and marginal" (Klápště, 2012).

Another question is whether the interpretative model of servant settlements could be transferred to the Czech history of the eleventh to twelfth centuries without a certain, if only implicitly considered, concept of the "state". "It was undoubtedly the latter", Krzemienska and Třeštík (1964) write of the servant organization, "which gave the ruler great economic opportunities and which thus ensured the stability of the Bohemian state for centuries to come." The service settlements were interpretatively anchored by the fact that, as we could read above, their justification was found "in the economic needs of the early medieval state". The landowning power is then attributed with a certain purposeful intentionality of their creation. Above we read that the prince (deliberately, purposefully) "settled [...] servants and craftsmen on the land around the courts and forts". However, the source reports from the early (Czech) Middle Ages do not tell us anything directly about such a thing. Krzemienska and Třeštík's interpretation is only functional if it includes the concept of a "state", which also tends to be a coherent "economic system". This, however, is an a priori research assumption. It probably reflects the optimistic confidence (not only Marxist-backed) of the intellectual classes of the 1960s in the social and economic functions of the state. This confidence may account for the anachronistic projection into the distant past of the concept of a strong state capable of maintaining rational economic purposes.

In addition to his conclusion pertaining to slave labor on the large estates, Třeštík also agrees with Šusta regarding the slave trade; though Třeštík depicts this trade more dramatically in terms of its scale and even its cruelty. In an essay from the late 1990s, Třeštík presents the Boleslavs' Czech state (tenth century) as "ex definitione Christian", but one that benefited greatly from the slave trade. This required "large-scale wars to be fought in the region to accumulate human booty, and at the same time it had to be offered in one place if possible, so that it was conveniently accessible to the merchants" (Třeštík, 1999). That place was the Prague market. Třeštík estimates from fragmentary source reports on slave prices how much profit the merchants could have made buying slaves (presumably captured pagans) in Bohemia and selling them in Islamic marketplaces. Třeštík assumes a considerable volume of this slave trade. (Krzemienska & Třeštík, 1964).

To summarize: the "servant settlement" or "system of servant settlements" can, in view of all that has been said, be considered—like the "large estate" above—to a large extent as an *ideal type*, or perhaps, since their image is reached by a certain inductive generalization, as Simmel's *social form*. The very formulation of this "ideal type" or this "social form", including its application as an interpretative tool, certainly took place in the field of historical science. In the same way, the conception of the deliberate slave trade for profit and the deliberate, even premeditated, creation of a network of servant organizations is also of the nature of an *ideal type*, and the idea of multiple profits from the sale of slaves is of the nature of a rhetorical plot within a *narrative interpretation*. Yet perhaps this too can still be situated on historical–scientific ground. However, the motives for *choosing* this particular line of interpretation were probably, as in the case of the "large estates" and the use of "slave labor" on them (see above), extra–scientific.

7.7 Back to "General History"?

The aim of this chapter has not been to draw conclusions about the existence and nature of slavery in Bohemia in the tenth to twelfth centuries. Rather, it concerns three facts. First: each specific community models the image of its past (Koselleck would say that it structures the "space of experience") according to its shared values, ideals and hopes (according to what lies upon its "horizon of expectation"). This is also manifested within the grounds of history. If we are historians living in a community romantically dreaming a naive dream of Slavic reciprocity, or if we are historians living in a community freshly finding hope in its contribution to the progress of Western civilization, or if we are historians living in an age that trusts in the organizational capacities of the state—essentially the same set of source traces would differ to us in legibility and meaning each time. At the same time, we can see how, in the thinking of some historians in the late nineteenth and early twentieth centuries, almost causally evolutionary concepts (one could say "historicist" concepts) were promoted and how these were gradually tempered by approaches corresponding to the theory of cultural diffusion (which were, in fact, already infused with the spirit of "historicism").

The abovementioned facts finally allow us to comment on the question of the possibility, necessity, and legitimacy of "general history". History, especially in the last three or four decades after the so–called "cultural turn" (Štaif, 2006), has tended towards microhistorical analysis (as practiced by Ginzburg, Le Goff, Duffy and Gurevich) and historical anthropology (Burke, 1997, 2019). This reinforces the conception of cultural relativism and the interest in individual agency, in the individual interpretative processes of semiosis as they had taken place in the minds of individual historical actors, in a part of history. The emphasis on the singularity and equivalence of cultures and the attempt to "return actors to history" then entails an underestimation, if not rejection, of general history.

In 1993, Richard van Dülmen, one of the prominent promoters of historical anthropology, judged that—since history and societies are constituted by individual and group social action—history is based neither on "objective" structures (and let us add: processes) as such nor their transformation, but only on their "subjective" perception: social reality is constituted through appropriation by individuals, groups and classes. If we are to succeed in our efforts to reconstruct power, economic and cultural structures, we must analyze them in the context of the relationship of the people acting through them and in the context of the relationships of these people to each other as they unfolded in everyday life (van Dülmen, 1993). In the early 1990s, these words were an innovative critique against the (objectivist) process and structure-oriented historiography of the 1960s to 1980s, practiced either in the French Annales school, in the German Historische Sozialwissenschaft, and finally within the framework of Middle and Eastern European Marxism. Today, thirty years later, these words would still sound too objectivist for some historians and theorists of historical sciences. For how to *reconstruct*, and what *objective structures*? Nevertheless, let us stick to Dülmen here. Building on his words above, we can say that in order to study the actorial ways of perceiving structures (and their processual transformations) as well as them being constituted by the actions of actors, we must be able to thematize these structures and the networks of relations within their frameworks. One of the ways is by comparing them on the ground of general history. And just to reiterate what was said in the introduction of this chapter, "general history" is not necessarily "general" history, and the goal of general history is not necessarily generalization, but can include recognition of the unique particularity of individual historical formations.

At this point we do not wish to decide definitively on the question of general history. We will note, however, that our discussions of slavery, the large estate, and servant settlements all show that we are on the ground of general history far more often than some theorists of history are willing to admit. And if this is so, then we should pay explicit attention to the theory of general history, just as biology pays attention to various forms of evolutionary theory.

References

Ankersmit, F. R. (1994). Six theses on narrativist philosophy of history. In *History and tropology. The rise and fall of metaphor* (pp. 33–43). University of California Press.

Barfield, O. (1973). *[1928] Poetic diction. A study of meaning*. Wesleyan University Press.

Budilová, L. J. (2019). *Od krevní msty k postsocialismu. Vývoj antropologického zájmu o Balkán. [From vendetta to post-socialism]*. Centrum pro studium demokracie a kultury.

Burke, P. (1997). *Varieties of cultural history*. Polity Press.

Burke, P. (2019). *What is Cultural History?* Polity Press.

Dopsch, A. (1909). *Die ältere Sozial– und Wirtschaftsferfassung der Alpenslaven [An older social and economic system in Alpine Slaves]*. Böhlau.

Foucault, M. (1990). *The History of Sexuality*. Vintage.

Ginzburg, C. (2013). *The Night Battles: Witchcraft and Agrarian Cults in the Sixteenth and Seventeenth Centuries*. Johns Hopkins University Press.

Gould, S. J. (1989). *Wonderful Life: The Burgess Shale and the Nature of History*. Norton & Co.

Havelka, M. (2019). Existují obecné dějiny? [General history – does it exist?]. *Dějiny—teorie—kritika, 16*, 264–269.

Horský, J. (2019). "Obecné dějiny", "obsahová filozofie dějin" a evoluční teorie. [General history]. *Dějiny—teorie—kritika, 16*, 253–263.
Hroch, M. (2016). *Hledání souvislostí. Eseje z komparativních dějin Evropy*. [Essays from Comparative history of Europe]. SLON.
Hroch, M. (2019). Obecné, světové, globální dějiny? [General or global history?]. *Dějiny–teorie–kritika, 16*, 270–274.
Jireček, H. (1863). *Slovanské právo v Čechách a na Moravě. Doba nejstarší: Od prvních zpráv do konce X. století [The Slavic law in Bohemia and Moravia in oldest times]* (Vol. vi, p. 237). Sklad Karla Bellmanna.
Kadlec, K. (1898). *Rodinný nedíl, čili zádruha v právu slovanském. [Zadruga in Slavic legal system]*. Česká akademie císaře Františka Josefa pro vědy, slovesnost a umění.
Kim, D.-Y. (2002). *Georg Simmel und Max Weber. Über zwei Entwicklungswege der Soziologie*. Opladen, Laske & Budrich.
Klápště, J. (2012). *Proměny českých zemí ve středověku. [Evolution of Czech Lands in the Middle Ages]*. Nakladatelstí Lidové noviny.
Koselleck, R. (2004). *[1979] Futures past: on the semantics of historical time*. Columbia University Press.
Krzemienska, B., & Třeštík, D. (1964). Služebná organizace v raně středověkých Čechách [The servant organization in Aerly Middle Ages' Bohemia]. *Československý časopis historický, 12*, 637–667.
Laslett, P., & Wall, R. (Eds.). (1972). *Household and Family in Past Time*. Cambridge University Press.
Martin, J., & Nitschke, A. (Eds.). (1986). *Zur Sozialgeschichte der Kindheit, Band 1,2*. Karl Alber.
Martin, J., & Zoeppfel, R. (Eds.). (1989). *Aufgaben, Rollen udn Räume von Frau und Mann, Band 1, 2*. Karl Alber.
Maur, E. (2001). *Gutsherrschaft und "zweite Leibeigenschaft" in Böhmen. Studien zur Wirtschafts–, Sozial– und Bevölkerungsgeschichte (14.–18. Jahrhundert) [Serfdom in Bohemia in 14–18th centuries]*. R. Oldenbourg.
Merhautová, A., & Třeštík, D. (1984). *Románské umění v Čechách a na Moravě [The Romanesque art in Bohemia and Moravia]*. Odeon.
Mitterauer, M. (1990). *Historisch–anthropologische Familienforschung. Fragestellungen und Zugangsweisen Wien*. Böhlau Verlag.
Müller, H. P., & Sigmund, S. (Eds.). (2014). *Max Weber Handbuch. Leben – Werk – Wirkung*. J. B. Metzler.
Peisker, J. (1899). *Slovo o zádruze [On zadruga]*. Fr. Řivnáč.
Pekař, J. (1900). K sporu o zádruhu staroslovanskou. [A dispute concerning zadruga]. *Český časopis historický, 6*, 243–267.
Pekař, J. (1901). K českým dějinám agrárním ve středověku. [The Czech history in the Middle Ages]. *Český časopis historický, 7*, 326–363.
Slavík, J. (1940). *Český národ v starším středověku [The Czech nation in the early Middle Ages]*. Pokrok.
Štaif, J. (2006). Sociální a kulturní dějiny. Několik poznámek k postmoderní diskusi [Social and cultural history]. In T. Borovský et al. (Eds.), *Kultura jako téma a problém dějepisectví (Země a kultura ve střední Evropě 2)* (pp. 35–47). Historický ústav AV ČR.
Šusta, J. (1899). Otroctví a velkostatek v Čechách [Slavery and farm estates in Bohemis]. *Český časopis historický, 5*, 34–43. a 86–97.
Todorova, M. (2001). On the epistemoloical value of family models: The Balkans within European pattern. In R. Wall et al. (Eds.), *Family history revisited comparative perspectives* (pp. 242–256). University of Delavare Press.
Třeštík, D. (1999). Kdo zaplatil vznik našeho státu? Říše prvních Boleslavů a otroci. [Who paid for the founding of ou State?]. *Dějiny a současnost, 21*(3), 2–7.
Troeltsch, E. (2024). *[1924] Historicism and its problems*. Fortress Press.

van Dülmen, R. (1993). *Gesellschaft der frühen Neuzeit: Kulturellen Handels und sozialer Prozess*. Böhlau.
Vašíček, Z. (1994). *L'archéologie, l'histoire, le passé: chapitres sur la présentation, l'épistémologie et l'ontologie du temps*. Kron.
von Below, G. (1928). *Die Entstehung der Soziologie?* Gustav Fischer.
Weber, M. (1904). Die "Objektivität" sozialwissenschaftlicher und sozialpolitischer Erkenntnis. *Arch Sozialwissenschaft und Sozialpolitik, 19*, 22–87.
Weber, M. (1991). In M. Sukale (Ed.), *Schriften zur Wissenschaftslehre*. Reclam.
Wocel, J. E. (1861). *O staročeském dědickém právu [The old Czech inheritance law]*. Pojednání KČSN. část V, svazek XI.

Chapter 8
Invariable Yet Changing

Abstract Biochemistry is an approach to describe timeless (ahistorical) functioning of life, yet it presupposes the extension of chemistry represented by ready–made biocatalysts, stored genetic information, and structures: the "logic" of the living. Problems arise also with determinism and its solution by allowing mutations. Moreover, codes and semiosis as basic achievements of evolution also reach out beyond chemistry. Reduction of life to genetic script as a basic level of description. Brief description of the standard model DNA, RNA, proteins and structures—copies, codes, interpretations. Dualism genotype—phenotype, and phenotype—Appearance. Underdetermination of biological data. Three forms of history attributed by science: beginning of the universe, second law of thermodynamics, and biological evolution. Attempts of scientific explanation of biological evolution, and conflict between Darwinian and Continental approaches. Neodarwinian synthesis as a scientific solution; mutation and selection. Microorganisms as an evolutionary model: evolution in the laboratory.

Keywords Universal laws · Science and history · "Logic" of the living · Genetic script · Reduction–inflation model · Evolution in laboratory

> [T]here is and will remain a Platonic element in science which could not be taken away without ruining it. Among the infinite diversity of singular phenomena science can only look for invariants. (Jacques Monod, 1972, p. 100)

Probably one of the greatest breakthroughs of the science of Modernity (i.e., since seventeenth century) was the assumption that events and constants are independent of the time and place in which they occur: that the laws found are valid at all times, and anywhere in the universe. If so, hypotheses can be repeatedly tested, and the results should provide identical or similar general patterns. In contrast, knowledge accumulated by the humanities that is strongly tied to time and place (and language) should be banished from (natural) science once and for all—and indeed the

A. Markoš, J. Horský, *The Seeds of Forms and the Forms of Seeds*, Biosemiotics 12, https://doi.org/10.1007/978-3-032-16131-4_8

seventeenth century saw a split between the two forms of knowledge, the coinoscopic and the idioscopic (see Glossary). Thanks to this development, both branches of knowledge got rid of many ballast and obsolete deposits from previous models of reality, often handed down since antiquity. Specifically, this involved the rejection of transcendent powers influencing the course of events: the axioms of the religious and faith-based paradigms were no longer to be considered in descriptions of the world and its history. (It is unnecessary to point out that this greatly simplified the situation; but new ballast may also have been created, and their residues may still influence *our* times.)

One alternative to rejection of history from science was the idea that its course is cyclical. Almost until the middle of the nineteenth century the axiom "the present is the key to the past" could therefore apply in historical sciences, including geology, astronomy and the evolution of life. As a consequence, the natural sciences (including biology) began to assume that life does not actually "exist" and can be left out from considerations: this so-called "mechanistic" view of life and its evolution became accepted. So too, timelessness of natural laws implies the rejection of any *goals* of cosmic functioning, as seen in the Aristotelian *causa finalis* and other forms of teleology.

Another novelty introduced by modernity is the assumption that whatever happens to an entity in nature must be explained by the action of forces *external* to it (Huygens, Newton): entities subjected to such forces thus became entirely passive, mortified "objects". This image of the world meant, among other things, the loss of its spirituality: for example, Kepler replaces the original concept of the soul of the world, the *anima mundi*, residing in the sun and moving through the planets, with the concept of force, *vis*. This trend has probably culminated in today's description of the three basic interactions (strong, electro–weak and gravitational) that, like Tolkiennian rings, hold the world together.

As a result of what was said above, three historical trends can be distinguished vaguely in the timelessness of Modern science—and must be swept aside. These are:

(1) Efforts to discover immutable laws *in the background* of all apparent dynamics that manifest in the (macroscopic) world; the realm of atoms and molecules are usually taken for such a background;
(2) Similar efforts that acknowledge the role of contingency; and
(3) Efforts that allow the world to shape its own history in abundant hylozoic interpretations.

The first trend—and after Darwin also the second—allows for applying methods of "normal science", i.e., it allows science to dwell in the realm of the Platonic world (see Monod's epigraph). The third approach—freed of hylozoic connotation—would acknowledge the innerness of living beings, with abilities such as introducing and interpreting the world via script, inventing novelties, hypolepsis, communication through codes and/or languages, etc.

Let's start our inquiry with a simple question.

8.1 How the Molecules Do It?

Biochemistry textbooks allow demonstrations of the *scientific* concept of the functioning of the living. Biochemistry does not deny evolution—but it hardly needs it: it seeks to discover laws as timeless as those used by chemistry. Yet it must allow quite odd and important extensions of chemistry: (1) the presence of biocatalysts (proteins) and the production of these catalysts with the help of (2) stored genetic script; (3) established codes; and last but not least, (4) the presence of pre-existing structures (patterns) and processes, as, e.g., metabolic pathways, membranes, etc.

All this has been a product of evolution. However, evolutionary inventions can be taken by biochemistry as ready–made and unchanging in real time; thanks to which it can observe their *chemical* behavior and explain their function. If possible, the properties of catalysts and processes will be tested and proven in vitro; and today, with the help of artificial intelligence, artificial catalysts can be tailored to desired new functions that are not naturally obtained in living beings (e.g., Polizzi & DeGrado, 2020; Lovelock et al., 2022). However complicated, they can be described by known chemical and physical rules—as the products of human technology can be described in just this way—there is no need to address the question of where they came from. We will again note that there is another kind of reductive understanding in our models of the world which can be summarized as follows: just as the elementary particles (a finite number of basic elements) build the diversity of the universe, just as matter (with the help of a finite number of kinds of atoms) organizes itself into a plethora of compounds, so too several groups of organic compounds underlie all the diversity of life, its "logic" and its seemingly apparent "disobedience" to ordinary physical principles.

Everything in the world is therefore seen through the lens of building "from below", from the "grassroots level". For the sake of honesty, we should point out that this idealized worldview is *literally* held by very few people today, it is a caricature; but let's stick to it for this moment.

The second question concerns time. Since the laws of physics and chemistry apply always and everywhere, physical time can be defined simply as the "ticking" that divides the course of physical processes into equal sections; the process being studied is itself deterministic or computable (at least statistically), and often even allows the arrow of time to be turned backwards without any consequence. Then, of course, history must either be excluded, or at least suppressed to a tolerable degree. In the second case, it must be assessed that history is also a lawful and deterministic process, just opaque to us (so far?) because of its complexity. Three such historical processes have somehow entered science in a disguise and must, somehow be dealt with (Sect. 8.3). Yet, the question arises again and again whether differing versions of world history should persist in science at all. After all, the main precaution was already established by Descartes: the object of investigation *must not* exhibit subjectivity. If we observe this condition, everything is the result of deterministic laws, i.e., the interaction between *objects* as worked out by physics, we just can't prove it "yet". Jacques Monod (1972), as shown in the epigraph, introduces a similar safeguard when he says that there is and will remain an element of Platonism in science, the removal of which would cause science to collapse. Within the infinite variety of unique phenomena, science cannot look for anything but invariants. For the lack of such invariants and laws, the deterministic (i.e., scientific) conceptions of human history (Hegel, Marx) have been swept out of science exactly because such prerequisites should be kept strictly. Only biological evolution was accepted in biological *science*, so below we must further examine the fact.

And even if we could put history aside, two questions arise concerning more than just life: is it at all possible to translate well and truly from one level of description to another? After all, the reduction of *things* into one kind of description, initially perceived as a refinement, as a better way understanding, can *explode* (in the sense of Lotman, see Sect. 4.4) into unexpected new dimensions; but it can also begin to blur in the stream of everyday communicative contexts (see the reduction-inflation model, Markoš & Cvrčková, 2013).

8.2 Genetic Script

The script is a paradigmatic example of such reduction of the world, be it the text of a human language or the genetic script of life. In both cases, there is a need of a reader—a living being—who is an interpreter of the message. Hence, before we move further, let's take a quick glimpse at the way how some parameters of the living are recorded in a medium that is *not* that parameter.

The *genetic script* is a non-random sequence of elementary components (called *bases*) in a linear polymer of nucleic acids (DNA or RNA). The ability of a cell to store and pass on to the next generation information about how to build a body (cells or Closure 2 communities such as a multicellular body) is usually conveyed by the slogan that *all* of the genetic notation of a cell (organism), i.e., "what is inherited", is present in the form of a non-random sequence of four *bases* in a DNA molecule. For the moment, let us just state dryly that this is an extreme simplification (reduction) of the type "all the world's misery is the fault of person (or ethnic group) XY". Of course, there are enormous advantages to limiting attention to DNA: *this* kind of information can be copied easily and with great precision, the sequence of its base can be quickly and reliably determined ("read") not only by cells, but also in the lab. In the second case, it can be converted into a linear sequence of characters, for example, on paper in the form of a sequence of *digital* markers, letters (e.g., ACGT), or by binary notation (0, 1). Such an inscription can be again (with possible modifications inserted by the experimenter) "embodied", i.e., transformed to cellular structures and monitored to see if some (artificial) mutation in the sequence will result in a change in the body phenotype.

Such a nucleic acid string can be handled, for example, as a program, a database or even a text, and this allows one to extract a huge amount of information. Or should we call it a "message"? This is because the technical term "information"

(Shannon & Weaver, 1949) has undergone inflation into common language and thus leads to misunderstandings. Shannonian information is a quantity of a certain value, and that value carries just and only *this* form of "meaning"; it is not concerned with any message contained within it. On this level, the genetic record (DNA) can even be said to be truly unambiguous, digital. This is further supported by another fact: although absolute digitality cannot be achieved in the real world, in the case of DNA the *discreteness* of bases ("characters") has been pushed to the utmost extreme.

In the context of genetic transcription and indeterminacy (see the next section), it is appropriate to note another remarkable dualism, that of phenotype–Appearance (see Sect. 4.3). Appearance, or form (habitus, likeness, *eidos*, *Gestalt*) refers to the overall aspects of a living creature as co–shaped by its memory and experience, and of both genetic and environmental influences. It may differ strongly at different stages of ontogeny (e.g., the embryo, caterpillar and adult butterfly). Unlike genotype, which can be "read" objectively as a sequence of discrete units, Appearance cannot be defined in this way. It is almost impossible to compare scientifically the different Appearances of individuals of a given species, which is why it is necessary to reduce Appearance, i.e., to objectify it to a *phenotype*, to well–selected (definable) traits that can be measured, recorded, or compared between individuals in a population. Such traits include, e.g., height, hair color, IQ, blood pressure, tendency to diabetes, etc.

An analogy with the "small world of drama" (Rosenberg, 1965) comes to mind: the "phenotypes" of the characters are given, but they often have little in common with the Appearance of full-blooded beings, i.e., actors; many of their characteristics are not mentioned at all. Thus, if phenotypic traits should be compared with some part of the genotype, or some structure, body function, age cohort, environment, etc., all such phenotypic criteria must be strictly monitored. An example of such a control may serve male mice of the same genotype (a completely artificial, laboratory–prepared, inbred population), in an age cohort of 6–12 months, taken from their mothers after 1 week, reared on a defined diet, five individuals per cage, etc. Even small details of the genetic composition and life-history of animals can affect some phenotypic traits under such defined conditions. They can thus reveal to the experimenter subtle details of genetic, metabolic, and other processes. Where such strict criteria cannot be met, as in a human—or virtually any natural population, a demanding statistical evaluation of the collected data comes into play. In historiography, the parallel of the pair Appearance–phenotype (or phenotypic set) is the pair "case–variables".

8.3 Underdetermination

Ankersmit's term, underdetermination, questions the effort of science to reduce things in the world to the smallest possible constituents, preferably describable by mathematical or logical formulas (see also Sect. 2.3. and this chapter's epigraph). The moment science ends up with formulas, it usually *must again step out of this area and give an interpretation of what they actually mean*; here again we encounter reduction followed by an inflation of interpretations. A good example of this is the wonderfully simple formula for defining entropy and the trouble of interpreting it—even after nearly two centuries. In biology, models of the Central Dogma of molecular biology or the mutation–selection model of evolution provide beautifully simple examples as well.

Methods of natural sciences are used throughout the whole range of research devoted to the analysis of the past, and even to the search for sources. Dating, authenticity, the relative age of findings, and their succession in time, the determination and measurement of proxy variables, etc., link all investigations of the past: cosmology, planetology, geology, climatology, paleontology, archaeology, forensic practices, and human history. There is no doubt about the relevance of these sciences for both biology and the humanities, but the question remains as to the *relevance* of *scientific* data obtained, for subsequent interpretation, explication, or guesses concerning past events (see the previous chapter). This is particularly difficult in the case of living beings (of course, we, authors, make no discovery in this respect). This is because scientific as well as historical data are much more accurate if one manages to keep—after Descartes—the subjective aspects of living beings to a minimum, as if they were interchangeable objects. This allows discovering subtle regularities of behavior (for example, in economics, evolution, sociology, or the formation of languages, nations and other social groups); one must only avoid absolutizing the findings and thus falling into "idolatry" concerning identification of "facts", concepts, or statements of authority.

With the advent of the Enlightenment, efforts were made to fit even historical sciences into the context of timeless science; cosmology, geology, the evolution of life, etc. were still in their infancy. Under the influence of the French Revolution, the idea of a knowable *goal* that we are approaching through *progress*, crept into all these descriptions. A goal, however, no longer revealed, but *scientifically* ascertainable and predictable—one almost wants to say "calculatable"—based on the claim that history is moving along cognizable trajectories. This trend culminated in the mid-nineteenth century with Hegel and Marx, as if building on the earlier chiliastic "history of creation" and its teleological component. These doctrines still operate today in various "disputes over the meaning of history" (Bultmann, 2019), and above all as ideology imprinted on the general consciousness of "common people" and politicians.

In the ninteenth to twentieth centuries, natural science—timeless, as we have seen—nevertheless takes three historical processes under its wing: the cosmology of the Big Bang, the second law of thermodynamics, and biological evolution as presented by Darwin. Geology, paleontology, archaeology, political economy, and other sciences have also sought inspiration from history.

(1) The first of historical processes attributed by, and compatible with science, is cosmological and concerns an unimaginably short period of time after the Big Bang. Then the fundamental constants and laws of the universe were set, at which point this historical process ended. Processes that took place in one way or another (although they could have taken place in other ways) during this extremely short period of time caused us to have this universe as we experience

it—and not another one. When that moment ended, however, cosmology became deterministic: at least it is not imagined that processes in different parts of the universe would have evolved divergently such that, say, an atom of iron would have different properties somewhere in the distance than it has here. We will not pay attention to *that* history.

(2) The second historical process is driven by the second principle of thermodynamics and represents a hypothesis (or axiom?) concerning how cosmogony will end one "day". The process is deterministic (or rather, teleological); it proclaims that all forms of "free energy" tend to turn into heat and dissipate in the universe and that all the structuration of the universe tends to homogenize. The process is identified as the tendency to maximizing entropy; when it reaches its maximum, evolution ends in timelessness. History enters this seemingly unambiguous, deterministic, and seamless degradation of the universe, only because the paths of decay are innumerable, proceeding at different speeds, intertwining, crossing, and otherwise influencing each other in different ways such that it is impossible to say exactly which way the degenerative process will go and how long it will last. It is even impossible to say even just for conditions on our planet.

This second path is accompanied by another process, which leads to the formation of macroscopic structures such as vortices, flames, stars, crystals, etc. These so–called *dissipative structures* seemingly resist the degradation process. They arise in systems that find themselves in a suitable steep of free-energy potential, and build formations *more* distant from thermodynamic equilibrium than the initial, more homogeneous arrangement. If a pile of leaves burns, the flame is more ordered (exhibiting lower entropy) than is the same pile. Yes, dissipative structures seemingly escape the action of the second law of thermodynamics, but they must "earn" it: they dissipate the energetic potential into the environment much faster (by burning) than if the gradient would dissipate spontaneously (by rotting). In other words, such structures catalyze the decay of the gradient, and "in return" they can use some of this energy in further self-structuration, to become even a better catalyst.

After the extinction of the gradient of energetic potential (to which they themselves contribute) dissipative structures either cease to exist (air currents, flame) or freeze in the form of a classical structure, such as a non-growing crystal or an extinct volcano. Self-structuring, so long as it persists, is a historical process: the direction of the dissipative structure's behavior cannot be predicted, nor can this trajectory be accurately reconstructed *ex post*. What is important, and what distinguishes dissipative structures from the history of the living, is that the components that organize themselves into a dissipative structure *do not have any innerness* (just as in the previous case the universe does not have it). They are drawn by purely physical processes; their description is admittedly much more complex than that of the "ordinary" tendency toward equilibrium but it is *physics* nonetheless. And in each instance, they arise *de novo*: a freshly started flame does not and cannot have the experience and memory of past combustions. (For a more detailed account, see, for example, Prigogine & Stengers, 1977, 1985).

Some properties of organisms (and even the biosphere as a whole) *can* undoubtedly be modeled as dissipative structures, just as other properties of organisms can be modeled as mechanical devices, chemical reactors or a computer. Indeed, both the biosphere and civilization significantly affect the dissipation of solar and telluric energy and build themselves up at the expense of such a process. Life itself, however, is *not* definable as a chemical reaction or dissipative structure.

(3) Finally, the third type of history involves living beings only: including the evolution of lineages, of the biosphere and of human cultures. Information, codes, memory, interpretation, established habits and experience all come into play with life, and through all these, living beings are able to *decide to behave* in one way or another. The reader already knows from previous chapters the two forms of historiography involved in relation to evolution. The first is spun by evolutionary biologists. These scholars may be regarded as a special variety of historians who describe not only the history of the human species but of all forms of life. However, in this case and in order to strengthen the scientific nature of the evolutionary story (as opposed to the human history), the notion of natural selection is introduced and declared to be a physical force; this seems to be how Darwin thought (Depew & Weber, 1996; see Sect. 4.5 and also below). However, we argue that in addition to evolutionary biologists and historiographers, living beings themselves can also act as "historiographers" and interpret their being in the world. They confront their momentary experience with stored memory (see hypolepsis) and *decide* (or *negotiate*) the trajectory into their future. With this, however, subjectivity would enter science, and it does not belong there! This is yet more confusing because evolutionary biologists are living beings themselves, and for the most part they don't question their own subjectivity. And this is why science, as it seeks to bring the evolution of life under its wing (while resigning to the history of the human species), has looked for ways to demonstrate that life is *also* ultimately a purely physical–chemical process. After all, life is implemented in matter, why wouldn't its components behave as matter? The assumption then is that the subjectivity (innerness) of the living is superfluous and even harmful for a real scientific description: it is misleading. Living beings are therefore not admitted as interpreters of their own evolution, life is only passively dragged through. In other words, living beings must—for the purposes of science—be reduced to objects. We do not intend merely criticizing this reductive model-building; after all, one of us is a biologist. We simply mantain that life is more than a biological *object*. Object is a model, constructed in a way to comply scientific demends. We discuss such scientific models of evolution below and in Chap. 9.

Hence, strandled between natural science and history, biology has made inroads in both directions. From its beginnings at the dawn of nineteenth century, it defined itself as a rigorous science—perhaps because, like physics, it needed to break away from natural philosophy, Aristotelianism, and various varieties of vitalism, Renaissance, Romantic ideas, etc. Life cycles, embryogenesis and organogenesis could be described as repeated developmental processes, and so there was no problem with their scientific description. Together with physiology, these movements sought to generalize and reveal the 'essence' of the processes described: as in other sciences, reduction to the general laws of living forms was their mission, though they did not always seek to find entirely scientific principles behind them. However, by attempting to describe the history of life and, above all, to explain its causes, evolutionary biology increasingly became the historical part of biology—and therefore a constant problem. Again, descriptive explanations have abounded since the eighteenth century, and biology often slipped into teleology to demonstrate that the evolution—just as ontogeny—is deterministic. Ever since Darwin... —but did his explanation strictly reflect these principles of natural sciences? We will address this question below.

8.4 The Scientific Character of Evolutionary Biology

As described by David Depew and Bruce Weber (1996), Darwin was an heir to Newtonian physics with its ethos of the explainability of the world; he therefore rejected teleological and vitalist versions of evolutionary theories of his peers, as well as the idea of cycles. Like the political economists before him (Adam Smith, Thomas R. Malthus), he found a solution in the analogy (analogy, not identity!) with gravity, the "invisible hand" that holds together large numbers of particles and imposes order on them. For Darwin, this invisible hand is natural selection, and it is (at least in principle) a *physical, quantifiable force*. And although evolution is a historical process, thanks to this, evolution belongs to science (that said, we do not know whether Darwin himself formulated it explicitly in this way; but see Sect. 4.5). For many researchers, this argument represents a major and fundamental distinction from historiography which, as we have seen, does not assume a similar, background physical force and works only with sources (traces) and their interpretation. However, biological evolution is also a contingent historical process, so even if we accept the argument that natural selection is a physical force, it is almost impossible to test hypotheses or conduct long–term experiments. Even *ex post* interpretation of the course of evolution is not easy: it allows for, to put it bluntly, an infinite number of theories, or an infinite number of interpretations. Below we will describe several attempts in this direction—as well as attempts to interpret them.

It is *this* variant of scientific version of evolutionary biology as conceived by Darwin that does not seem to have been understood on the Continent: of course, scientists on continental Europe were also seeking a *scientific* explanation. Early in the twentieth century in his *History of Biological Theories* (1930),[1] Emanuel Radl subjects Darwin's theory of evolution to an extensive analysis and concludes that Darwinism is dead because biologists on the continent did not find it *sufficiently* (yes: sufficiently) scientific:

[1] This English edition is an abridged version of a 2-volume German treatise from 1905 and 1909.

> They were all awaiting new light on the question of development [read: evolution], and behold! it came from Darwin; but it came in a manner new to science. He offered neither morphological argument nor physiological experiment nor profound speculation concerning the nature of life, but, in the place of these, an enormous mass of quotations from books of travel. He himself belonged to the hated materialists, and deprived the older thinkers of their dearest belief—their belief that Nature is rational. Is it surprising that so much obloquy was heaped on the new theory and its exponents? (Radl, 1930, p. 56).

On the belief in the intelligibility of nature we give just two then contemporary quotations that were, and still are, repeated *ad nauseam*: Theodor Eimer (1897, p. xi) writes: “No one rational will deny that everything in nature must, of course, obey laws. But the changes in various directions promoted by Darwin are disordered and random—as opposed to lawful, proven development according to determinate rules.” And the vitalist Hans Driesch (1929, p. 4) joins in: “The ideal of which I speak is the ‘law of nature’, and this ideal is nothing less than one of the postulates which make science possible at all.” They are right.

To quote Radl again: “The aim of the new theory was to describe the natural agency that led to the origin of species (to *describe*, not to *understand*), to adhere to nature with reason so that our narrative would be a repetition of what happens in it. Darwin asks for the opposite of what his predecessors wanted: instead of grasping the diversity of nature in general terms, he asks that reason dissolve itself in the particulars of things. Previously, reason (ideas, noumena) had been regarded as the eternal truth proper, while phenomena had been regarded as the fleeting image of reality. Darwin does just the opposite: he does not pay any attention to reason and concepts, and the course of events itself is everything to him. Everything that looks like a concept, an abstraction, a logic, is rejected” (from the Czech edition, Rádl 2006, p. 118; this paragraph is not contained in the English translation.).

In Radl’s time, the Continent triumphs: with the advent of genetics, biology begins to approach science, and Darwin is pushed aside. Only the so–called Modern Synthesis in the 1930s links Darwinism with genetics, and Neo-Darwinism emerges, of which we are still heirs today.

Zdeněk Neubauer (2010), in the remarkable essay, *Darwin’s Discovery of Nature as it can be read from Radl*, provides another interpretation of the events after Darwin’s getting out:

“In the spirit of the liberal tradition, Darwin meant the expression *Laws of Nature* (*leges naturae*) as the genitive *subiecti*—not *obiecti*, as it has been understood until now and as Darwin was and still is interpreted. It is therefore not laws *forcing* nature to obey, but laws naturally established by nature itself. Nature has thus been elevated for the first time to the subject of legislation in the natural sense, where ‘laws’ arise of their own accord—with tradition. And they change with application and with interpretation, the two—interpretation and decree—falling into one. Moreover, natural laws are also natural in the sense that they do not exist supernaturally as if valid *in abstracto*, but are applied *in concreto* to the here-and-now of each individual, in individual cases of procreation and survival, just as in society laws exist only in their application in legal cases and legal decrees/decisions" (Neubauer, 2010, p. 292).[2]

Neubauer believes that Darwin offered a kind of—unrecognized by him—resurrection of natural philosophy, with a new recognition of the spontaneity and creativity of nature unbounded by the laws imposed by science. He points to the enthusiastic reception of *Origin* by the lay public of the Continent, and, on the contrary, to the harsh rejection by both "dull" modern science and the ruling conservative circles.

Throughout our book, we seek to make the reader accept our position that Darwinian evolution is *primarily* a historical process that falls not under natural science but under historiography. We attempt to demonstrate this below; but let us try to outline three main assumptions of biology as a science.

The first, a priori assumption of the scientific status of biology is the view that living beings are objects, and as such they can, and must, be denied their innerness. What is sought are laws, physical forces like any other, which move an object; the idea that the object *acts* cannot be accepted, objects do not act. As has been repeated since Aristotle:

"Nature does nothing in vain but always what is best from among the possibilities for the substance of each kind of animal, which is why if it is best in a certain way, it is also in this way according to nature" (*De incessu animalium* 2, 704b15–18). And another quote:

"Nature does everything either because it is (conditionally) necessary or because it is better" (*Generatio animalium* I, 4, 717a15–6).

Radl's contemporary Jakob von Uexküll (e.g., 2010) does acknowledge the innerness of an animal, yet he would promote views quite similar to those of Aristotle. His texts are full of depictions of natural events as a well-played counterpoint, harmony, meaning. "Nothing is left to chance in nature" (Uexküll, 2010, p. 100). When a butterfly scares away a bird by spreading its wings, "the butterfly does not know that the sparrow flees at the sight of a cat's eyes. However, that which brings this umwelt-composition into being exhibits an awareness of these facts" (Uexküll, 2010, p. 102). The "that which brings" is not specified in any way: is it nature (as in Aristotle), is it God, or perhaps Latour's OWWAAB (see Chap. 1)? Everything happens like a massive and well-rehearsed symphony of nature (or, elsewhere, a glockenspiel), and the task of biology is to discover the score by which it is played; there is no roam for contingency or even chance. The whole biosphere works like a mutual tuning of the "players". This approach understandably provoked controversy with Uexküll's contemporaries—biologists who lived within a different scientific paradigm and believed in evolution. Uexküll argues instead that this mighty symphony has been played within *every* period of evolution, albeit with understandable differences, with different musicians and instruments. Life has a 'meaning score' such that, for example, the spider builds its web because it is 'fly-sensitive' (*fliegehaft*), it has an idea of a counterpoint that needs to be played relative to the fly. Also, the cell has its own "personal tone" that matches the tones of the other cells in the organism, it is aware of the meaning of what is being played.

---

[2] Bruno Latour's opinion is quite similar, see Chap. 1.

Box
Of bubbles and foams is the title of a paper by Markoš & Švorcová (2024); it is an attempt to absorb the idea of the immutable and perfectly tuned *umwelt* bubble by the contemporary movement in the biospheric, evolutionary thinking. The authors consider the "bubble" as stuffed of historical memory and experience, of which the particular individual (or species) makes use only of a negligible fraction. The rest of this umwelt-pool "waits" for its chance to enter again into the whirl of events—by chance, by anamnesis, or in case of a coincidence of events. Thanks to the pool, the particular bubble has many overlaps with other *umwelten* thus forming an ever-changing "foam", i.e., the biosphere. Not a symphony, but a jam session.

Even today, some researchers deny such a status to plants and "lower" animals, not speaking of bacteria and archea. And the scheme begins to fall apart even with mammals and birds and: though we are animals by constitution, our species simply also cannot fit within it. The discomfort that pre-Darwinian biologists and evolutionists had with this situation is readily apparent in their writings.

Let us note here that the advent of bacteriology facilitated ways out of such clench. This new science was unencumbered by tradition, and it is much easier to treat microorganisms as objects than, say, elephants. At the same time, much of what was discovered about microbes could be extrapolated (in terms of scientific interpretation) to macroscopic forms of life: hence the famous slogan saying that "What is true for a bacterium is also true for an elephant".

Second, even if we take living beings for biological objects, they have—even according to a schedule that denies them an innerness—a quality incomparable to the objects of physics. They reproduce (*not* in the manner of a chain reaction known from chemistry and physics), thus giving rise to similar (but not identical) "objects", often through complicated ontogenetic processes and growth. Each lineage necessarily carries a specific kind of memory, such that the descendants of, say, an oak tree will be oaks, not ducks or even maple trees. If we accept that life has an innerness and that living things can therefore *make decisions*, our task becomes even harder, because adhering to the scientific basis of the experiment will require a drastic reduction in the *umwelt* of individuals in the population and of the population as a whole: only in this way can the decision-making process be minimized.

Third, both life and its reproduction require drawing resources from the environment, and that resources are not unlimited and so there is competition for them. This is where the classic role of natural selection comes in.

As an adherent of Newtonian physics, Darwin had no difficulty with the first point: that is why his theory was and is called "materialistic" (meant pejoratively). He also emphasized the similarity but not identicality of offspring in reproduction; and he formulated one of the pillars of his theory—*descent with modification*. The emphasis on resources (taken from Malthus) then inspired him to coin the phrase *struggle for existence*, a competition—among those dissimilar offspring—for resources in which the winner is the one who not only survives but is able to produce offspring.

Life is simply the ability to produce offspring—as many offspring as possible. This ability, as we have seen, is then evaluated by an invisible hand—*natural selection*—which determines the trade-offs (but it must not be called a negotiation) between the environment and all "living objects" and declares the *fitness of* these objects. Finally, since Darwin was studying multicellular organisms, the main criterion for judging the qualities of individual living things was their Appearance and behavior.

8.5 Present

Let us skip over the waves of various Darwinisms and focus on the contemporary Neo–Darwinism of the so–called Modern synthesis, which has been taking shape since the 1930s. This is a synthesis because it is a combination of genetics and Darwinism; it introduced advanced statistical methods to assess genotype-phenotype relationships, not only at the level of individuals but of entire populations. The Modern synthesis became completed as a coherent theory in the 1970s, hand in hand with the key discoveries in molecular biology.

By the 1970s it was known that proteins are powerful regulatory and structural molecules, that they are synthesized as linear, non-monotonic polymers of 20 *amino acids*, and that the order of amino acids in a polymer determines the reliability of its functioning; substituting a single acid in a single position can—even drastically—alter this functioning. It was also known that the instructions on how to put amino acids in sequence are stored in *genes*. It turned out (after a long period of various speculations) that a gene can also be viewed as a linear polymer of four *bases* (in DNA, sometimes RNA). Again, the substitution of a single base *can* alter the order of amino acids in a protein and thus its function; all the genes in the body then form its genotype. At this point, we need only state the axiom: *genes determine phenotype*, and we no longer need pay attention to the phenotypes (let alone Appearances); we need only compare the genes or genotypes of individual organisms. The heterogeneity of alleles can be measured—and we enter a field of objectivity with much more certainty than when we determine and compare the properties of phenotypes. And it is more "precise": for example, two varieties of apple tree may differ in a single identified allele or even in a single nucleotide, whereas the differences in the *macroscopic* phenotype may not be that clear: what criterion can we use to distinguish them—the size of the apples, their taste, their resistance to pests?

When studying evolution, we cannot overlook factors caused by contingency. Gould (1989) drew attention to the matter with his metaphor of "rewinding the tape": if you rewind the "tape of evolution" to the beginning and let it play again, it is quite certain that each time it plays different: often even unforeseen life forms may appear in the "output". Certainly, the analogy is not very successful: if you rewind and replay a tape of a Beatles recording, you will not expect to hear a piano concerto. (The analogy has caught on, however, and now it presents a different problem: younger generations no longer know what a cassette player was.)

The history of this controversy is stormy and we will not discuss it here; however, it is clear that in the study of evolution as a contingent history it is particularly advantageous to use populations with short lifespans that are adapted to laboratory conditions: on such populations (as we will see below in Sect. 8.6), evolution can be traced on a set of parallel samples. We can even freeze the samples (aliquots) and later "revive" them and subject them to further evolution. We will be able to identify mutations—even hidden ones that may only become apparent after further mutations have occurred.

8.6 Evolution of Bacteria in Solitary Confinement

In experiments of this sort, a population of organisms is allowed to grow in several parallel cultures under controlled conditions, and the evolution of selected parameters in each parallel are thoroughly monitored. The initial and intermediate conditions are either kept constant as far as possible in all parallel samples, or they may be slightly modified according to the experimental protocol. Although these are laboratory experiments, the results can shed light on the possible role of contingency in the natural world.

Microorganisms (bacteria, archaea and protists that can be grown in a laboratory) are particularly suitable for such experiments because they are easy to handle, have short generation times and large populations, and their (typically) asexual propagation allows the establishment of parallel populations with identical clonal genotypes. In addition, some species can be frozen and later revived, allowing the preservation of "frozen fossils" from different phases of the experiment.

There are bacterial species that can produce ten or more generations per day under defined conditions, and thus the evolution of the relationship between genome mutation and cell phenotype can be observed under strictly regulated external conditions:

(1) When we grow a bacterial culture under certain conditions, we get a *strain* with characteristic properties that distinguish it from other strains of the same species (for example, one that does not cause diarrhea in humans and can therefore be safely handled in the laboratory). From a single cell of the strain, we can then grow a *clone*, a population that is the offspring of a single cell, so we can assume that—at the beginning of the experiment, the cells in the population are identical. (2) Preference is given to clones that cannot differentiate into different cell types (e.g., they are unable to produce spores); (3) no contact with other biosphere inhabitants is allowed, and (4) the population can be grown in suspension in defined, shaken nutrient media.[3] In such conditions, the population is *reduced to a single phenotypic*

[3] The defined nutrient medium is prepared only from well-defined constituents. This fact is emphasized here because some cell cultures require for their growth a medium supplemented with blood serum (usually calf serum), which contains a large number of unknown components in varying concentrations; the composition of the medium may therefore vary from batch to batch.

object—with the environment comparable to that of a prisoner in solitary confinement. However, there *is* one escape: mutation. But even this is usually not dependent on the activity of the bacteria, it is the work of chance. (5) Perhaps there is a mutation-free finesse: an evolutionarily stable strategy, a "division of labor" within the population, whereby differentiation into multiple cell types occurs.

The bacterium *Escherichia coli* is frequently used in such "evolution in a test tube" studies. In this state, a clone can be maintained for tens of thousands of generations—and an experimenter can easily immerse himself in such experiments within the timeframe of her/his career.

The metabolic vigor of the clone determines how well the bacteria will multiply in the medium offered: the rate of multiplication is a measurable criterion of *fitness*, an indicator of how efficiently they convert the resources offered by the medium into "living matter". Some cells of a clone may mutate, for example, in such a way that they make better use of the resources offered, and thereby multiply faster and eventually dominate the population. This purely biochemical phenotype is likely to correspond to the genotype—parallel clones can be compared in terms of the mutational distance of certain sections of genomes or even whole genomes (again, a quantifiable criterion) during modelled evolution. Mutations that do not manifest themselves biochemically in this way have no effect on *fitness* and thus usually represent a "null phenotype". However, even if they do not manifest phenotypically, they may persist in the population and reveal their effect in concert with another mutation if it occurs later (see analysis of the Cit^+ phenotype below).

Below, we present a summary of a long-term experiment investigating the role of contingency in the evolution of bacterial population. The experiment was initiated in 1988 by R. Lenski, and the monitoring continues to this day; the example discussed here (Blount et al., 2018) is from work in his lab and reports on the evolution of an *E. coli* clone over 60,000 generations. In 2022, after 75 thousand generations, Lenski passed the baton to his successor (*Nature* 606, 634 presents an interview with experimenters).

Initial conditions (the starting point in the adaptive landscape) are determined by the characterization of the chosen bacterial clone; with boundary conditions prescribed by confinement in a culture vessel, in a defined and shaken medium. This means that the four billion years of previous evolution are ignored; for a clone, all cells can be assumed to be identical—except for mutants that arise later. The experiment is run in twelve parallel populations of the clone, each of which is daily transferred into fresh medium (details see below), and selected parameters are also measured daily; aliquots of the population become frozen at regular intervals, so that later—even years later—it is possible to "go back in time", "read" the genomes and possibly study their evolution further. (In the case of human history, a similar example would be a revival of a medieval city or Caesar's military company.)

To prevent the bacteria from multiplying into a high density (that subsequently stops growth), the culture must be *passaged* regularly. In the experiment reported here, this is done by taking a small portion (an aliquot) of the grown suspension once a day[4] and diluting it approximately 100 times with a new medium. "Approximately" because the newly started populations in *all* parallel cultures must always be the same size at the beginning of the next daily cycle, with no regard to whether or not they have grown to different densities (depending on their *fitness*) in the previous round. Thus, the only variables are (i) the cell density at the time of the next sampling (fitness level), (ii) cyclic daily changes in the medium due to metabolism, and (iii) genome mutations (these can also be detected retrospectively in this case, by DNA sequencing from stored aliquots). It is assumed that the only memory in the population is DNA, i.e., the genetic script, and that this will be read by the cell in the manner of a computer and not a reader. The "historical sources" are therefore of a single kind and are equally valid for all.

During a one-day period, the cells can produce about seven generations, i.e., the population density increases about 120 times, the concentration of the main nutrient—glucose—in the medium probably decreases drastically during that time, and metabolic end-products will accumulate in the medium. The conditions in the culture are therefore continuously changing, but cyclically return to the baseline every day—thanks to passaging.

What to expect from the experiment:

(1) In most cases, the parallel populations will change only slightly during the experiment (and thus differ only slightly in fitness) and the changes in their genetic signatures will follow similar, but not necessarily identical, trends (mutations may occur at different locations in the genome). However, only those individuals in each population that—however narrowly—win the competition for the only available nutrient source, glucose, will begin to dominate. This is because there is only one possible adaptive peak (see Glossary: Adaptive Landscape), and in all parallels, it will represent the highest efficiency of use of *the same* energy and organic matter source, i.e. glucose (although other possible sources are present in the medium, but these cannot be processed as nutrients by the population under study). Those cells in the population which, due to contingent mutations in genes related to glucose metabolism, increase this efficiency even slightly, will multiply faster (expressing increased fitness) and will thus dominate the population over time. Because the experiment is long term, *all* parallel populations will reach a higher, but approximately equal, fitness, and may reach it by different mutational pathways. There are multiple paths to the top.

[4]The passaging mode is important for another reason. Although the bacteria would persist in the sample for days and months, they would soon cease to be a uniform clone. Depletion of the nutrients originally present in the medium leads, among other things, to *actively induced* increased mutagenesis (the so-called SOS reaction) "in the hope" that it might lead to a solution to the problem in at least a small fraction of cells. Most cells naturally die as a result of such gambling, but some find a solution to process the dead cells' material and survive. Long-term preservation of a clone (or its "historical stages") is therefore usually only possible by suspending life processes—freezing the cells at very low temperatures.

(2) The population is divided into two or more subpopulations, each of which specializes in some part of the processes leading to higher fitness: the sample ceases to be homogeneous. Since the overall fitness of such a heterogeneous population depends on the presence of all the variants, an *evolutionarily stable strategy* is established where all subpopulations will propagate in equal proportions. Again, the most likely trajectory will lead to a fitness peak associated with glucose utilization. However, the community may not simply be the result of a genome mutation: the frequent occurrences of the division of labor suggest a differentiation process rather than parallel genetic evolution.
(3) Rarely, mutations occur that allow the use of another substance in the medium, a substance that the cells could not metabolize before. Below is a case of a rare mutation Cit^+.

Blount et al. (2018) found that after 31,000 generations one population developed the ability to grow aerobically on citrate (Cit+); citrate was part of the medium from the very beginning, but not as a nutrient. So, the ability to grow on citrate is highly beneficial, yet the Cit + trait has only appeared in one of the 12 populations. It could have been a single, extremely rare mutation—and it could (just as easily and at any time) have appeared in any other populations. However, it could also have been a result of several mutations in train: A, B and C must have appeared first, but did not affect fitness in any way. They were, however, a precondition for the subsequent mutation D to have an effect: only then did the corresponding Cit^+ phenotype appear. This was indeed the case: the phenotype would be the result of a unique history during which one or more mutations accumulated, and subsequently potentiated the resulting appearance of the trait.

References

Blount, Z. D., Lenski, R. E., Losos, J., & B. (2018). Contingency and determinism in evolution: replaying life's tape. *Science, 362*, eaam5979.

Bultmann, R. (2019). *History and Eschatology: The Presence of Eternity*. Baylor University Press.

Depew, D. J., & Weber, B. H. (1996). *Darwinism evolving. Systems dynamics and the genealogy of natural selection*. MIT Press.

Driesch, H. (1929). *[1907, 1908] The Science and Philosophy of the Organism* (2nd ed.). A. & C. Black.

Eimer, G. H. T. (1897). *Orthogenesis der Schmetterlinge: Ein Beweis bestimmt berichteter Entwickelung und Ohnmacht der natürlichen Zuchtwahl bei der Artbildung; zugleich eine Erwiderung an August Weismann [Orthogenesis in butterflies]*. Engelmann.

Gould, S. J. (1989). *Wonderful Life: The Burgess Shale and the Nature of History*. Norton & Co.

Lovelock, S. L., et al. (2022). The road to fully programmable protein catalysis. *Nature, 606*, 49–58.

Markoš, A., & Cvrčková, F. (2013). The meaning(s) of information, code … and meaning. *Biosemiotics, 6*, 61–75.

Markoš, A., & Švorcová, J. (2024). Of bubbles and foams: Umwelt counterpoints in symbiosis. *Sign System Studies, 52*, 438–456.

Monod, J. (1972). *[1970] Chance and necessity. An essay on the natural philosophy of modern biology*. W. Collins Sons & Co.

Neubauer, Z. (2010). Darwinův objev přirozenosti, jak jej lze vyčíst z Rádla [Darwin's Discovery of Nature as it can be read from Radl]. In A. Markoš (Ed.), *Jazyková metafora živého* (pp. 285–329). Pavel Mervart.
Polizzi, N. F., & DeGrado, W. F. (2020). A defined structural unit enables de novo design of small–molecule–binding proteins. *Science, 369*, 1227–1233.
Prigogine, I., & Stengers, I. (1977). The new alliance. *Scientia, 112*, 259–302. and 643-652.
Prigogine, I., & Stengers, I. (1985). *Order out of chaos*. Flamingo.
Radl, E. (1930). *The history of biological theories*. Oxford University Press.
Rádl, E. (2006). In T. Hermann et al. (Eds.), *Dějiny biologických teorií novověku II.: Dějiny evolučních teorií v biologii 19. století [History of biological teories in the Modernity, Vol. II]*. Academia.
Rosenberg, H. (1965). *The tradition of the new*. McGraw–Hill.
Shannon, C. E., & Weaver, W. (1949). *The mathematical theory of communication*. University of Illinois Press.
von Uexküll, J. (2010). [1940] The theory of meaning. In D. Favareau (Ed.), *Essential readings in biosemiotics. Anthology and commentary* (pp. 81–114). Springer.

Chapter 9
Not by Genes Alone

Abstract Evolution may be seen as a process moving between two limits: one is the Standard evolutionary theory (SET), with genes seen as programs that almost totally control the deeds of the body. The hypothetical opposite pole is Extended evolutionary theory (EET) thar would assume some hypothetical organisms that are able to tinker by unlimited interpretation of genetic texts being at their disposal. Real theories work with real organisms that move in the space between such extremes. Here we give some examples of activity within the field. The evolution of the species *Homo sapiens* is in the focus of many teams that hunt after the "genes of the soul" that would distinguish our species from other members from the genus *Homo*. Some point mutations in regulatory proteins (and perhaps even proteins involved in metabolism) may turn out to be culprits. Genetics may one day become replenished by epigenetic processes that work with much subtler regulatory power.

Other teams have their eyes on evolutionary processes like genetic assimilation, canalization towards some outputs, plasticity of species under various condition, etc. The chapter gives also examples of big adaptive radiations occurring in ecosystems (as contrasting with modelling evolution in laboratories), to illustrate achievements and difficulties of such studies.

Keywords Genetic assimilation · Evolution of *H. sapiens* · Epigenetics · Adaptive radiation · Plasticity · Evolution in situ

> Words whose meanings are relatively fixed and established, words which can be defined—words—words, that is, which are used with precise connotation by different speakers—are *results*, they are *things become*. The arrangement and re-arrangement of such univocal terms in a series of propositions is a function of *logic*, whose object is elucidation and elimination of error. [Continues in the footnote] The use of them is left to the logician, who, in his endeavor to keep them steady and thus fit them to his laws, is continually seeking to *reduce* their meaning. I say seeking to do so, because logic is essentially a compromise. He could only evolve a language, whose propositions would *really* obey the laws of thought only by eliminating meaning altogether. But he compromises before this zero-point is reached. Owen Barfield (1973), p. 131

A. Markoš, J. Horský, *The Seeds of Forms and the Forms of Seeds*, Biosemiotics 12, https://doi.org/10.1007/978-3-032-16131-4_9

For the title of this chapter, we borrowed first part of the book heading by Richerson and Boyd (2006); here, however, we have broadened its topic from human evolution to the evolution of all life. We will give examples of how the genetic script may be confronted with the larger world, be it established by tradition (what Kosseleck would call "collective memory"), by the momentary settings within the biosphere ("communicative memory"), and by ever-changing (often quite quickly) physical environment.

Since the 1940s, biological research in cell biology and evolution became increasingly "genocentric". Eventually, when the "gene" was identified with the sequence ("script") of discrete elements (bases) in DNA, the slogan took roots in the form "It's all in the genetic notation of DNA". In the 1970s, Dawkins (1976) expressed it succinctly with the analogy of the replicator and the vehicle: only the script of DNA can be copied (i.e., replicated) *very precisely*, and all the information for the construction of the body should be stored in it. The body then serves only as a DNA vehicle, a "copying machine" that carries the instructions stored in the script to build the next generation of vehicles. The DNA thus acts as a parasite (the "selfish gene") that has domesticated the body and has also "built" the body in its image. The phenotype of a cell and of a multicellular organism depends on the reading (execution of programs) of different parts of this transcript (more precisely, of variants of readings in different situations or at different stages of ontogeny). The script itself also determines the order and/or situations in which its parts are read. For example, although we carry many of the genes "for" embryogenesis in all our cells, they were only actively read (they "controlled the processes") only for a short time after fertilization. Their activity (expression) is triggered over time, either automatically in a proper timing and position cues, or it is switched on/off by the environment, as for example the melanin in the skin in "Caucasians" is formed by the action of ultraviolet radiation on the skin; similarly, wound healing programs "switch on" when the organism is damaged.

The decisive role of the script in evolution was generally accepted, and hereditary changes in "phenotypes" (cells, individuals of species) were understood to be due to its mutations, which were (with some exceptions) regarded as purely random, caused either by inaccurate copying or by external physical influences, for example, hard radiation, or certain substances, but not by the action of the "hardware", i.e., Closures of which the cells in question are a part. We emphasize that from this view the body itself and its interactions (what we have called Closures) are mere "products" of genes, fulfilling the role of a passive object (see Chap. 1). The subjectivity of living beings is not considered; all life falls into the same category as, for example, planets or stones. One certainly cannot know everything about things, but as much as is known, no things "care" about their properties within this paradigm. This is why self-care of living beings must be limited as much as possible when experimenting with something that is alive—as in the experiment with bacteria described in the previous chapter. Here we add some examples also from other life-forms that appear to counter the genocentric model.

9.1 Inheritance of Cell Structures

The inheritance of structures, which, like Lego blocks, consist of (or are composed by) products of genes and other substances, is in some cases peculiar because the newly assembled structures may not be able to arise spontaneously, but only based upon (or according to) the existing structure. Of course, the cell itself is an example of this, but let us focus here on some cellular structures. The kinetosome is an example of such a structure that is required in eukaryotic cells to build flagella, cilia, centrioles, and the structures derived thereof. Destruction of kinetosome(s) results in an inability of the cell to build such structures; despite this fact the genetic transcript remains unchanged and the cell is able to synthesize all the necessary components ("Lego parts") of the structure—only the directions how to assemble them into a functional unit is not at hand. The principle applies to higher order structures as well: in ciliates, a row of cilia (each quill growing out of a single kinetosome) can be (surgically) rotated 180 degrees, so that it "rows" against the efforts of the other rows, and this anatomical nonsense will be propagated for hundreds of generations (e.g. Sonneborn, 1950). What we have depicted in the example of the kinetosome is also true for some other structures of the cell: for example, in ciliates, the cell mouth is also assembled in this way. It seems that a kind of "crystallization nucleus" must also be present in some cell wall structures for those groups that have a wall. Above all, however, this applies to biological membranes, which apparently never form *de novo*: they grow only by insertion of newly synthesized subunits into the existing membrane (e.g., Reber et al., 2023).

In all such cases, *changes may become hereditary*, passed on to the offspring; if such an "epimutation" occurs in nature, natural selection may sometimes spare or even prefer such an epimutant. In the world of just-established Neo-Darwinism such information was not a very welcome finding; moreover, in the USSR Lysenkoism was rampant at that time and similar findings seemed to "play into its hands".

9.2 Genetic Assimilation

This case is much more complicated than the previous one, and the history of the discovery of this "anomaly" and its explanation is particularly instructive. The story starts with the mutants of the *Drosophila* fly: dozens or even hundreds of *genetic* mutations can be induced in the laboratory by mutagenic substances, and such mutations may give rise to a recognizable phenotype. In adults, phenotypes include, for example, deformed or differently shaped wings, wingless individuals or, on the contrary, double wings, limbs instead of antennae, different shades of eye color, etc. Such mutations are compatible with life (under laboratory conditions), and the mutations are, of course, inherited in subsequent generations (i.e. they are either phenotypically manifested, or the alleles are kept in a latent state and manifest only under certain statistically mapped circumstances).

In the 1950s and 1960s, Conrad Waddington exposed the *larvae* of fruit flies (*Drosophila*) to various harmful substances (e.g. ether vapors), which led to anomalous phenotypes in mature flies (Waddington, 1975). However, these turned up to be only different types of anatomical mutilations, not mutations: the affected flies therefore produced normal offspring. Waddington went on to select some well-recognized "cripples", crossed them among themselves, and re-exposed the (otherwise normally developing) larvae of their offspring to the same noxious exposure as their parents. He repeated this for several (roughly 10) generations—then it turned out that the crippled phenotypes were also *stabilized genetically,* without further toxic exposure. Thus, Waddington acquired heritable changes *without* inducing random genetic mutations. The finding, termed *genetic assimilation*, seemed to grossly contradict the *Central dogma of molecular biology* (Crick, 1958; see the Glossary) and caused embarrassment that persisted into the late 1990s.

In order to understand this phenomenon, we must first mention the existence of *chaperonins* or "heat shock" proteins (*hsp*): they are supposed to help native proteins to take the correct shape needed for them to function properly. Of course, *all* multiprotein complexes in the cell do this, but here we will focus only on special cases requiring the action of the *hsp*. Some native proteins cannot do without them at all, but we will be interested here in those chaperonins that appear only in crisis situations and act as a kind of prostheses for damaged proteins: damaged either by mutation, by heavy metals, toxic agents, or high temperatures (where they were first detected, hence the name "heat shock"). They impose on them—albeit mutilated—a standard shape and function. However, under excessive stress—mutational or toxic—the "repair" capacity of the *hsp* proteins may not be sufficient and the phenotype will be disturbed anyway, primarily as the result of malfunction of some proteins. This apparently occurred after repeated crosses between Waddington's "cripples"; but we still need another clue.

A second clue to understanding the observed phenomenon is the fact that phenotypic traits are often caused not by a single, but by a number of different proteins. For example, suppose that the "notched wing" phenotype is the result of the activity of eight proteins, let's call them A–H. Suppose further that if two of them are mutated or damaged (say, C and H, or B and D), the phenotypic changes do not occur, the *hsp* proteins are still able to fix them. However, if four mutated alleles meet in a single individual through crossing the "cripples", the *hsp* capacity may no longer be sufficient. Suppose that in the population of fruit flies that

Waddington took into the experiment, such mutations of the B, C, D and H genes occurred at low frequency, and scattered among many individuals: the probability that all four mutations would normally come together in a single cross was small. However, Waddington's exposure to toxic substances and subsequent selective crosses gave a noticeable shock to the organism's regulatory systems, and hsp suddenly ran around like headless chickens. Epimutants began to appear, not receiving "repairs" any more: in such cases, external toxic exposure caused additional stress over that already present due to the presence of mutant alleles, and as a result, altered phenotypes appeared. Obviously, the intercrossing increased the likelihood that all the mutant alleles of the A-H complex would accumulate in the offspring over time: if it happened, there was no need anymore for external boosts, and a stable and heritable mutant phenotype would emerge after a few generations, even *without external action* and without inducing gene mutations. It is therefore not genetic assimilation, but the concentration of several defective alleles into a single organism by crossing followed by selection; this could also be achieved by simple crossbreeding, but would take much longer. Susan Lindquist and her group (Rutherford & Lindquist, 1998, and a series of papers that followed from her lab) provided this elegant explanation as part of a much broader investigation of the function of chaperonins, proteins that can help cells resist exposure to harmful agents for long periods of time, and can also mask harmful mutations that only become phenotypically apparent as a result of the coincidence, in certain circumstances.

9.3 Signatures of "Humanization"

The zoologist and anthropologist Adolf Portmann (1897–1982) in the 1940s theorized over the evolution of humans. He, of course, was not the first to state that the human newborn differs substantially from newborns of other great apes. Whereas the latter can see immediately, hold themselves to their mothers, respond to their surroundings, and their brain development (in terms of size) is completed, the human brain continues to grow for a year after birth, and many basic functions and skills are acquired even later, during the long years of childhood and adolescence. Portmann's explanation lies in the "premature fetus" (compared to great apes, human gestation should take up to 15 months): a larger brain needs a larger cranium, but the larger head of a "mature" fetus would then no longer pass through the birth canals. These, in turn, cannot be adequately enlarged for anatomical reasons, i.e. because humans walk on two, and a wider canal would make it difficult if not impossible (at least for women) to walk. Up to this point Portmann, we further draw mainly from the textbook on human evolution by Burda et al. (2014).

Natural selection has thus directed our evolution towards "premature" births. (Leaving aside, for example, the other apomorphies of the genus *Homo*, the evolution of the hand and the brain; more on this can be found in the textbook mentioned above.) But what was the cause and what was the effect? The Hominina group (which includes, among others, the genera *Australopithecus*, *Paranthropus*, *Ardipithecus*, and later several species of the genus *Homo*) split off from the related great apes some 7.5 million years ago, and is characterized *entirely* by bipedal walking. Their young, however, were apparently born with smaller heads and brains; it wasn't until long afterwards, some quarter of a million years ago, that species of the genus *Homo* "felt the need" for a larger brain. At this point, even when the different programming of ontogeny led to "premature birth", nothing actually happened, even such a child could be raised and raising became a habit, eventually fixed genetically (see above genetic assimilation). Or was it rather a complex and long-term sequence of mutations selected—as the hypotheses below show? The archaeological sources (or is it still paleontology?) allow a following story: 2.6 million years ago, at the onset of the chains of glacial periods, our ancestors (*H. erectus*, *H. heidelbergensis*) showed the ability to use fire and perhaps also tools—unmistakable evidence that "something" had started to happen.

As a *pars pro toto* of the further interpretation, let us quote the following: "By the mid-twentieth century most scientists had agreed that human bodies and minds were the product of genes that had evolved under the pressures of natural selection, just like everything else in the living world" (Hayden, 2009, p. 776). As the author of the paper goes on, scientists believed that identifying the genes that distinguish us from other creatures would help reveal the cause of human apomorphies, especially, of course, speech and mind. In fact, the twentieth and twenty-first centuries have only shown, as usual, how complicated things are, and attention is also increasingly focused on the cultural performance of human society; evolutionary anthropology seeks to synthesize both perspectives. Still, some genes deserve attention; below, then, are some reflections on this topic.

9.3.1 The Hunt for the Genes of the Soul

If we stick to the idea that there is a basic level of description at the level of genetic script from which the human Appearance derives (and is primarily explicable), we must necessarily look for those parts of the genetic notation that are found only in humans. In line with such a reasoning, *Homo sapiens* must *necessarily be* genetically distinct from sister species such as Neanderthals or Denisovans (older hominin forms would certainly come in handy as well, but their DNA is not—at least not yet—available). And given that we have the entire genome of the Neanderthals and a significant part of the genome of the Denisovans at hand, we need no longer limit ourselves to some hypothetical "proto–human", or "apeman" defined on the basis of anatomical (phenotypic) data, i.e. basically on comparisons of the volume of the cranium or the shape of teeth and bones.

The problem is that the genomes of the three species of humans—and also contemporary great apes—are (despite the separation of the two lineages, we recall, only 7.5 million years ago) largely similar (in terms of the number of genes and the organization of the entire genome).[1] Subtle differences found between them are very difficult to explain as the very *cause* of differences in Appearance (or in phenotypic traits). Thus, if we want to explain them purely genetically, they would have to lie either in some physiological mutation, or in differences in regulation. Below, we refer three results of such a painstaking search pointing to three candidates (probably more genes could be found, but not many more). For a better understanding, it should be said that some proteins (and their genes) are represented in the organism (not only in humans, of course) as a family of closely related variants (but not alleles; see Glossary for the difference), i.e., homologous genes, so-called *paralogs*, or proteins encoded by them. The same (or perhaps almost the same) function is often performed in different types of cells or tissues by different members of such a protein family—i.e., according to the expression of this-or-that gene, the cells of the tissues carry a kind of signature derived from the position of organ primordia in the early embryo. For example, the histologically unrecognizable muscles of, say, the thumb and little finger, or the trunk and head, show differences in the expression of such paralogs, and these differences were already established in the basic plan ("map") of the new individual during embryogenesis (even before the primordia of the corresponding tissues and structures could be recognized). The authors of the stories presented below convince the reader that it is the function of some specific paralog that is *the* apomorphy "responsible" for our being here. Note also that these considerations do not include the mutations that have led to the variation in human populations—for example, the type of skin color, the ability to live at high altitudes (Tibet, the Andes), the ability to survive on a diet virtually devoid of sugars (the Inuit), or to digest seaweed polysaccharides (the Japanese). Despite such racial differences, what unites us all is the ability to think and speak conceptually, and these are traits that evolutionary anthropologists would have us believe other species of the genus *Homo* may have lacked. The focus, then, is on the "genes for mind" that perhaps distinguish us from other extinct hominins. Such genes promote the development of the cranium and cerebral cortex, and their mutation in modern humans would lead to anatomical changes (e.g. microcephaly) or "feeble-mindedness".

[1] So the well-known slogan "We are 98% chimps", or its reverse version "Chimps are 98% like us".

Box. Analogy from the Historiography

An example, this time from the history of Christianity, is the deep and centuries–long schism between the Eastern and Western Christian Churches. It is often found in two "point" mutations in the canonical texts—in a single letter of doctrine: was Christ *homoúsios* or *homoiúsios* (consubstantial with the Father, or proceeding from the Father)? The second such "mutation" of essentially the same kind is the conjunction "and" (Latin *-que* at the end of the word "son": *filioque*) inserted into the creed of Western Christianity. Formally, the schism occurred in 1054, and both sides of the dispute have been trying to prove that causes mentioned above were the *cause* of the schism ever since (see, e.g., Franzen, 1969; Romanidis, 1981; Tomkins, 2005). But perhaps this event is just an afterthought to justify the divergent development of the two great Christian civilizations. Then we would see an "epigenetic" change in the understanding of the world, additionally "assimilated" in the record. Whatever the case, the histories of Eastern and Latin civilizations took different directions.

We can use this example to compare the procedures of evolutionary biology and historical science. If we were to treat the sources, to use Vašíček's (1994) term, as "witnesses" and work by the so-called "direct method" we would actually be proceeding within the framework of history as evolutionary biology would proceed if it relied on point mutation. We would consider the "i" or "-que" as a direct witness of a substantial event, even as its fundamental factor and immediate cause of further development. Or we would attribute an unambiguous meaning to this "i" or "-que" (as the archivist Divíšek does with the different sizes of letters and numbers in the inscription, see Chap. 2). However, as historiography "archaeologizes" (Vašíček, 1994) and begins to treat sources as "indicators", it tends toward an interpretation that is analogous to the one labeled here as "epigenetic". The "i" or "-que" can then be understood as a (relatively arbitrary) means of articulating a variety of attitudes and feelings; they are a vehicle, a component of a sign, a symbol through which a broader complex of cultural contents (or differences between cultural complexes) or feelings of sympathy and antipathy are expressed. We can formulate a number of "theories" or "historical interpretations" (Ankersmit, 1994, see Sect. 2.3) —"possible histories" —about the nature of these complexes and feelings (and expose them to the "veto of the sources", cf. Koselleck, 2004). What Ankersmit says about the underdeterminacy of theories and interpretations, will be true about their relation to "data" and about the difference from evolutionary biology in terms of the possibility of establishing this relationship. Keep this example in reading following examples.

1. *Point mutation and the development of mind.* A single small change or substitution can have far-reaching and long-lasting impact on the subsequent history of a species by placing the mutant gene in new and unexpected contexts (see

"mondegreen", the charade with mutations of letter sequences or sounds). A point substitution of a single base ("letter") through an altered protein often "shoots" through all levels and alters the Appearance of the organism. These are inborn metabolic disorders such as sickle cell anemia, phenylketonuria, cystic fibrosis and dozens of similar diseases; to this are added changes caused polygenically, where it is not so easy to pinpoint the exact genetic "cause" (for example, many mental disorders). Do we owe our existence to some similar mutation(s)? Given the high similarity of hominid genomes mentioned above, the only way is to look for really small differences in base sequence—in genes and in regulatory regions on DNA; but there are thousands of these, hence, one must also find functional differences that will correlate with such substitution.

Pinson et al. (2022) discovered one such needle in a haystack, a mutation that may have had a significant, directly decisive effect on the phenotype of *Homo sapiens* compared to Neanderthals. It is the TKTL1 gene with a point mutation, causing the resulting protein in *H. sapiens* differ from both Neanderthals and Denisovans (and also from all other mammals, because with the exception of *H. sapiens* it is a highly conserved gene) in a single amino acid. Why did they start to investigate this "needle"?

The TKTL1 protein is one of the paralogues of the enzyme transketolase, a common housekeeping enzyme of which there are hundreds of variants in the cell. It is involved in the synthesis of fatty acids and steroids, as well as the fixation of CO_2. In humans and other mammals, the paralog TKTL1 is activated in the frontal lobe of the forebrain (neocortex) of the fetus; in other parts of the body, numerous other paralogs of the same enzyme do the job. The neocortex is most responsible for cognitive abilities in mammals, and the TKTL1 mutation in humans correlates with stimulation of one group of neural progenitor cells, so that these produce many more neurons in the neocortex in the human fetus than is normal in other mammals. The correlation of the "human" version of the gene with neocortex development and the associated abilities to think and speak has understandably intrigued evolutionary anthropologists. But can a trivial mutation of a common metabolic enzyme lead to the emergence of something as complex as "mental abilities"? Indeed, the authors explain such an effect "metabolically": more lipids are produced, and such an equipment allows for more intense cell division, etc. Of course, a more refined explanation is also offered, at the level of regulation. For a number of common metabolic enzymes, it has been shown that, in addition to their ability to control some functions, they can also bind to protein complexes in chromatin and thus influence the expression of various genes.

The cited paper by Pinson et al. also provides other data: the authors experimentally verified the effect of human mutation in several ways. It was relatively simple to show that the activity of the enzyme in the frontal lobe of the human fetus does indeed increase between 14–19 weeks of gestation, and not in other period of development and/or in other parts of the brain (this was, of course, studied on spontaneously aborted fetuses), and correlates with the rampant division of one group of neurogenic cells; the mouse homologue of the gene does not act this way in the mouse fetus. By gene manipulation, the authors therefore replaced the mouse TKTL1 gene with a human variant, and also with a variant present in Neanderthals: as expected, proliferation and differentiation of cells in the mouse (and also in the ferret) neocortex was stimulated only by the *H. sapiens* variant. Another approach involved experiments on organ cultures obtained from human fetuses aborted during the critical developmental period. In the control group, the relevant cells divided normally, but not in the group that had the gene removed ("knock-outed"). Similar results were obtained in organoids, this time grown in culture from human cells (see Sect. 10.4.3 for organoids): replacing the gene with the Neanderthal gene again slowed down the proliferation of the progenitors. Finally, inhibition of the metabolic pathways of which this particular transketolase paralog is a part, had a similar result (hence the "metabolic" interpretation mentioned above). Thus far, the effect of the mutation has been detected at the level of metabolism; however, the effect at the level of chromatin will be certainly discovered sooner or later.

Thus, we see that a difference of a single amino acid may throw the protein into a different region of the space of possibilities, and the mutant often gets into completely different contexts because of it.

2. *The theory of the copious "wiring"* of the regulatory gene FOXP2. The FOXP2 protein is one of a vast number of transcriptional regulators, i.e. proteins that alone or in concert with many others decide which proteins will or will not be synthesized in a given cell. The FOX family alone has 19 members (FOX A-S), and each family may have additional paralogs (FOXP has four); and there are dozens of similar regulatory families. The FOXP2 protein alone binds to dozens, perhaps hundreds, of regulatory sequences, and *somehow* influences the activity of the genes linked to them.

The human FOXP2 mutation was discovered in 2001 through a family history of three generations with severe heritable speech disorders. Since then, a flood of papers has been published describing various features of this "speech gene"; here we draw mainly on two review articles (den Hoed et al., 2021; Fisher, 2019). In addition to affecting motor skills of speech organs, mutations in the gene caused changes in brain anatomy and disruptions in the growth and intergrowth of nerve cells, leading to a range of mental disorders and autism; in animals, it has been linked to changes in vocalization in birds and bats. As in the previous case (TKTL1), the protein affects, among other things, the morphogenesis of certain parts of the brain, especially in the fetus.

On evolution: the FOXP2 of *Homo sapiens* and *H. neandertalensis differs* from its homologue in great apes in a single amino acid substitution in the protein. Probably, the theory goes, it influences the expression of more genes in the human brain than in great apes. Also, "humanized" mice with human FOXP2 show brain changes similar to those of humans. Hence, the mutated gene is probably the sole *cause* of the drastically altered phenotype. Similar findings then lead to the speculation that this mutation may have been the cause of the enormous brain evolution within the genus *Homo*.

3. *The theory of bad jaw clenching.* The protein myosin is responsible for the transformation of chemical energy into mechanical: it catalyzes all sorts of movements and changes in cell (and vicariously the whole organ) shape, and is naturally found in large quantities in muscles. There are about 16 paralogous genes in the human myosin family, many of which are expressed in striated muscle: in different muscles in different proportions and ratios; muscles, depending on their position in the body (limbs, masticatory muscles, etc.), prefer one or the other version of myosin. Great apes have well-developed masticatory muscles anchored to a correspondingly massive area on the skull. However, in the lineage leading to humans, the dominant version of the myosin gene typical for the chewing muscles underwent a severe gene disruption for this version of myosin sometime 2.5 million years ago (the time of the event was calculated by comparative methods from gene sequences), resulting in a non-functional protein—and vastly underdeveloped human chewing muscles compared to, say, a gorilla. We chew thanks to other versions of myosin in our chewing muscles, but we can't crush nuts or bones.

And now the story (Stedman et al., 2004): The poorly developed human masticatory muscles did not need a massive attachment on the skull any more, and this became lost over time. This freed the cranium from constraints—and it could start to grow according to the needs of the growing brain. As a consequence, human cognitive abilities increased: it may be no coincidence that it was 2.6 million years ago that people started using fire and roasting meat to compensate for their lack of mastication skills! Another benefit as that people warmed up—the glacial periods just started. A beautiful historical narration based on traces!

4. *New Gene Theory.* Even more intriguing may be the discovery of entirely new genes not found in related primates (hominoids)—such a discovery would suggest that we have become "hopeful monsters" that have embarked on the path to spirituality. First, again, a brief introduction. Transcription (of DNA into RNA) refers not only to protein-coding RNA (mRNA) and the auxiliary translational machinery (tRNAs and rRNAs). These classes of RNAs exit the nucleus into the cytoplasm and function there: they can leave the nucleus thanks to a short regulatory stretch of the molecule (an analogy to the post-office code). Most types of RNA (from other classes), however, do not leave the nucleus and participate in building the chromatin structure and regulating its functions. One class of these nuclear transcripts is composed of hundreds of lncRNAs (long non-coding, i.e. long chain, yet non-coding any protein; Marchese et al., 2017). Among these, An et al. (2023) searched for mutations such that would be found only in our species and that would, in addition, attach a routing "address" to the lncRNA chain, thus allowing it to leave the nucleus, and in the cytoplasm to became translated into a protein. Such a new protein could rebuild the network of interactions in certain cell types, and if it did so in cells of the future cerebral cortex (as in the case of the TKTL1 mutation above), it could even pave the way to "mentalization". Of 74 identified lncRNA exported from the nucleus, the authors went on to examine just one. What the function of the resulting protein is coded by it, authors do not

know yet. However, its manifestation involves accelerated growth of brain organoids (even after transfer to macaque organoids) and also (after transfer of the gene to mouse cells) quantitative growth of the mouse forebrain—manifestations similar to those of TKTL1 and FOXP2. The difference, however, is that this is a *new* protein, specific *only* to humans, while the first two represent only mutations of common proteins known also elsewhere. With the advent of artificial intelligence capable of analyzing large number of sequences, we are sure to see many more discoveries in this area.

9.3.2 Why Couldn't It Have Gone Faster?

Let us pause to address an agonizing question: what did our ancestors do for the tens of thousands of years that they seem to have done nothing? At least 100,000 years ago, humans were already anatomically—at least in terms of skeletal remains—indistinguishable from modern humans. We therefore can infer from this that their cognitive abilities also were on a par with ours. But why didn't we fly to the Moon 90,000 years ago? We have read somewhere the hypothesis that they looked like us, but they didn't talk, thus they needed another tens of thousand years for their vocal cords to mutate intensely, which seems like a far-fetched speculation. But did we have to mutate? Isn't this more likely to be "ecological" and associated "epigenetic" changes, i.e. pressures not from molecules but rather from some of the forms of Closure 2 change, such as the nature of human communities at the end of the ice ages and in the Holocene? Such considerations of "somatization" was already discussed in Sect. 6.6.3.

9.4 Adaptive Radiation

Here we approach a counterpart of bacteria in a solitary confinement (see Sect. 8.6), a full-fetched evolution of species in a natural environment. We will see the obstacles in such situations where initial and boundary conditions of experiment can be stated only vaguely (if at all) and the process itself is influenced by ecological interaction (often unknown or only vaguely stated).

First the definition, as we read it on the cover of Schluter's book *Ecology of adaptive radiation*: "Adaptive radiation is the evolution of diversity within a rapidly multiplying lineage. It occurs when a single ancestral species differentiates into an impressive array of species inhabiting a variety of environments. Much of life's diversity has arisen during adaptive radiations. Some of the most famous recent examples include the East African cichlid fishes, the Hawaiian silverswords and, of course, Darwin's Galapagos finches." (Schluter, 2000).

We further learn from the book that adaptive radiation is the result of competition for resources and (divergent) natural selection that will favor the key innovation, i.e. the expansion of the umwelt of the species. As the author notes, it is not easy to prove these theoretical assumptions in practice. Nor is it easy to distinguish adaptive radiations from non-adaptive, "just so" radiations, as in the case of invasive species. Importantly (Schluter, 2000, p. 49), the dominant ecological trend in the case of adaptive radiation is not so much specialization as it is continuous spread to new environments (followed by speciation subsequently). Particularly valuable are the cases of parallel adaptive radiation in two similar, isolated environments (again, we allude to the motif of "replying the tape" in a slightly different way—on several tape recorders simultaneously).

In the natural environment, the study of the evolutionary process on a short time scale—as opposed to laboratory-grown microorganisms—becomes much more difficult: rhe main reason being the generation time of the species under study and also because of their sexual reproduction. Also, the presence of humans and their interventions in the landscape greatly interfere with monitoring. One way to provide initial and boundary conditions, and to limit the number of environmental variables, is to study the penetration of individuals or small populations into a new environment in which the species has not lived before. Such a population will remain isolated from the parent population and begin to diversify. Ideally, the colonization of a new territory occurs only once, without subsequent waves of immigrants, so that the diversity of a single lineage originating from the founder population of the species can be determined.

Divergent evolution, and its criteria are summarized by Schluter (2000), p. 131f as follows:

(1) Phenotypic differences between populations and species should have a genetic basis; (2) Chance should be ruled out as an explanation of the pattern; (3) Population and species differences must represent evolutionary shifts and not just species sorting; (4) Shifts in resource use should match changes in morphology or other phenotypic traits; (5) Environmental differences between sites of sympatry and allopatry must be controlled; and (6) Independent evidence should be gained that similar phenotypes compete for resources.

Schluter acknowledges that only a small proportion of the adaptive radiations studied meet all six criteria (only 5 of the 94 cases cited; 23 meets at least four conditions), and he also admits that not all criteria need always be considered. Note that neither he nor other authors admit what we emphasize in this book: in their conception, evolution merely happens to lineages; creatures play no part in it, again except perhaps the role of sexual selection. Let's try to read this with the insight that living things could also forge their own way through the ecospace: after all, in their original homeland they also lived in a complex ecosystem—and they may have had plenty of records (memory traces) in their umwelt to help them navigate in the new environment. Anamnesis? This does not reduce the quality of purely biological studies, it's just a question of interpreting them, and the interpretation has to be supported by appropriate data collection, and it is not an easy task.

In particular, islands, isolated lakes, or high mountains provide best opportunities for adaptive radiation. Examples include the Galapagos or Hawaiian Islands, the Aleutians and many others: the oldest of such island chain emerged as an oceanic volcano (magma burst its way through the oceanic crust). The oceanic crust moved sideways, separated the new island and its volcano from its source, and magma therefore periodically made its way up from the depths to form another volcanic island. Thus, chains of unequally aged islands were and are formed. The terrestrial inhabitants got to the islands from the mainland, or from older islands.

A classic example of adaptive radiation is the "Darwin's finches" of the Galapagos (the oldest island is about 14 million years old), a group of birds whose ancestors were brought to the Galapagos from the mainland some one million years ago (Rubin et al., 2022; Schluter, 2000 states 3 millions); the descendants gradually colonized all the islands of the archipelago and specialized in different types of food—a fact that is visibly reflected especially in the shape of the beak. The Galapagos Islands are far from the mainland and further waves of colonization are unlikely to have occurred; hence, the species diversity of this group of birds probably comes only from the founding population (perhaps slightly "spoiled" by later inter–island migrations). On his return from his round-the-world voyage, Darwin studied the collected specimens of these birds, and this experience played a significant, if not decisive, role in the formulation of his theory. However, geographically defined islands are not the only habitats with adaptive radiation: similar processes can also be found in mutually isolated rivers and reservoirs, or in mountains.

If we want to study the adaptive radiation *currently underway*, a number of technical problems arise:

1. Animals studied are multicellular organisms, mostly with sexual reproduction. Inbreeding or at least crossbreeding within varieties cannot be avoided—as is the case with strains of laboratory or domesticated organisms, or with self-pollinating plants. Nor can be introduction of "invaders" from other populations living in neighboring territories or repeatedly coming from a source population prevented. The offspring are therefore heterogeneous within the entire population studied. Heterogeneity can be suppressed by the textbook ideal case of introducing a fertilized female to an island where no other members of her species are present: if she has young, she will establish a population there by inbreeding. The situations described are often cited as the ideal basis for the emergence of a new species, and even more interesting would be if such a homogeneous population started to diversify and fill (or establish) niches in the newly colonized environment.
2. The environment is variable. In the case of colonization of a new fresh-water reservoirs by ocean fishes, this may include differences between its bottom, water column and surface of a lake, constantly changing day length, fluctuating temperatures, local population of competing species, predators, food, etc. Ocean migrants have not had to deal with similar problems before. There are even more issues like this in the tropical forest.

3. Speciation does not have to be caused only by genetic reasons (random mutations of the genome): it can be a simple shift in the available space of possibilities, or epigenetic changes that either disappear after several generations or become fixed by genetic assimilation (see above). Consider the frequent differences between upland and lowland forms of plants, which, when taken into a different environment, adapt to it, even though it may take a few generations. Also, parallel adaptations need not be genetic; for example, we find trees in most plant families. The mutation theory (change in allele frequency) is still preferred because it is essentially irreversible and also because it can indeed often be correlated with changes in phenotype. Most importantly, in case of a mutation we can at least theoretically distinguish between speciation and simple adaptation within a species. Therefore, Schluter (2000, p. 22) will assume that "phenotypic differences between species have a genetic basis, despite the fact that this has not been confirmed in all cases". Most experimentalists enter research with this belief, while theoretical biologists point to a number of other factors, such as organismal plasticity (see, e.g., 9.5, Pfennig, 2021b).
4. The organisms under study usually produce only 1–2 generations per year (or more, but they live in hiding as, e.g., small mammals), so the experiments are very time and resource-consuming. Populations are also much less numerous than can be achieved with microorganisms.[2]
5. On the other hand, the possibility of studying the Appearance (habitus) of the organism, its ontogeny and physiology is an advantage—although often the research must be limited to a few phenotypic traits.
6. A special subject of research is the study of parallel evolution—the emergence of the same phenotypes in mutually isolated but similar habitats (see examples below).

Box. Disgression to History

In history we can encounter similar efforts, albeit in a much "softer" form. As discussed in Chap. 7, some scholars are now quite suspicious of such approaches to general comparative history, on the one hand, for ideological reasons (let us leave these aside, as they should not be given a say in the field of science), and on the other, for methodological reasons. Among the methodological objections to such procedures are both those already mentioned above, namely that it is quite difficult to identify unambiguously initial and boundary conditions for the cultural developmental lineages ("series") being compare. The problematic nature of declaring certain cultural variants occurring in different developmental lineages to be "the same" so as to be

[2] These are all constraints even in case of multicellular and sexual laboratory models: hence the preference for small and rapidly reproducing organisms such as the fruit fly (*Drosophila*), nematode (*Caenorhabditis*), tiny *Danio* fish, mice and rats, thale cress (*Arabidopsis*), etc.

analogous to the same phenotypes, is also a drawback. An example of such a comparison within world history is the theory of the so-called "axial age" or "axial transformation". According to this theory, in the four centers of cultural development (Greece, Israel, India, China), analogous transformations in the thematization of the cosmos, society and the individual were supposed to have occurred, essentially independently of each other in terms of their basic nature. This theory, formulated more than a century ago (Alfred Weber, Karl Jaspers), cannot—it goes without saying—be applied explicatively to individual texts or artefacts that have survived from a given region or period. Rather, it is a demonstration of how the West structured its cultural and historical identity a century ago. Another example might be studies comparing more than a hundred human languages and seeking to find a more general (but not universal) scheme of how the names for the six or eleven basic colors evolved in them (Berlin & Kay, 1969; Kay & Maffi, 2000; Kay & McDaniel, 1978). In these cases, the attempt is to uncover analogous developments while excluding the possibility of cultural transmission (cultural diffusion). In the case of the "urban revolutions" of the fourteenth and fifteenth centuries or the national movements of stateless nations in the nineteenth and twentieth centuries, cultural transmission must, however, be assumed (i.e. guild masters imitated each other in their revolt against the patriciate across Europe; national builders motivated each other).

Back to adaptive radiation. There are few suitable models and the results are often ambiguous; there are even fewer targeted experiments, i.e. deliberately man–made. Here we recall three cases, where the "experiment" was set in motion by nature itself: the fish of glacial lakes; the fish of East African lakes; and the adaptive radiation of small lizards of the genus *Anolis* from the American mainland to the Greater Antilles, and thence either back to the continent, or to the Lesser Antilles.

Glacial Lakes and Connections to the Sea Let's first look at the conquest of freshwater by the stickleback (*Gasterosteus acuelatus*), although it is far from being the only species on which the process of land conquest can be observed. The stickleback is a small fish of the northern hemisphere, originally oceanic. It is characterized by a number of conspicuous and locally variable morphological features such as spines, size, coloration, etc.—making the different forms easily distinguishable to the trained eye. The species has also repeatedly invaded brackish and fresh waters at different times, and this change of habitat has again resulted in a number of morphological divergences from the marine forms. We will be interested in one relatively recent event of such an incursion into fresh waters (there are also much older invasions documented, for example from the Mediterranean). When the glaciers began to melt at the end of the last glacial period 12,000 years ago, and at the same time ocean levels began to rise, a number of glacial lakes were formed on land (Canada, Finland, the Pacific coast), into which marine fish penetrated, adapted to fresh water and established separate populations, more or less isolated from each other, and also from the marine homewaters. This is also how sticklebacks penetrated freshwater. The fish from the isolated basins are distinguishable from each other, but they all share one peculiarity: in each lake they form two ecoforms, one benthic (living and feeding at the bottom) and one limnetic (from the water column). The two forms can interbreed within a lake, but normally do not, living separate ecologically and spatially (and experimental fish from different lakes will preferentially breed with their corresponding counterpart). The whole phenomenon seems to suggest that two very similar, perhaps even identical, subspecies occur *in parallel* in freshwater lakes. Or are they species? Multiple parallel "tape replays" regularly lead to the same result—are there perhaps some evolutionary trends? Or are we looking at a mere example of a plastic phase in evolution of a species or even a higher taxon? Bony scaled fish (Teleostei) are the youngest class of vertebrates (they appeared 200 million years ago, in the Triassic) and the diversity of this group is high. It may also be a plastic reaction, a shift of forms into new (and similar) successional lines within their umwelt—then what we observe could be "just" differences of the type between upland and lowland forms, as we repeatedly observe in the same plant species (see Sect. 9.5). Research on the genomes of forms from different lakes reveals a number of mutations, often of regulatory genes; whether these are the result of natural selection or random correlations is disputed.

Lakes of East Africa South-North across Africa (from Mosambique to the Red Sea) stretches the Great Rift, which marks the point where the continent will break up in the future and a new ocean will open up between the emerging land plates (currently the rift is 30–100 km wide and continues through the Red Sea and Jordan Valley to Syria—6000 km in total). On the African mainland, the rift has two arms that encircle part of the territory of Tanzania, Kenya and Uganda, only to merge into one again. The depression began to open about 35 million years ago[3] and is filled with deep lakes (up to 2000 m); they vary in age (e.g., Malawi two, Tanganyika ten million years) and are only partially connected to each other; they are coupled to the Indian and Atlantic Ocean basins as well as the Mediterranean Sea via outflowing rivers) On the plain between the two arms of the depression is Lake Victoria, relatively shallow (up to 80 m) and young (about 400 thousand years old). For further interpretation, it should be noted that this lake dried up a few thousand years ago[4] and did not refill until about 12,500 years ago.

The larger lakes are inhabited by a large number of species (500–2000) of cichlid fishes, and the spectrum of this richness has evolved locally, i.e. the species are not shared between lakes, although they are often indistinguishable in form (parallel evolution); in some lakes the entire richness even seems to be the offspring of a single pair. The fish come from marine populations that are not particularly diverse, and they probably came to the lakes by chance, unleashing a downright frenzied

---

[3] For comparison, the Atlantic Ocean has been stretching between Africa and South America for 140 million years.

[4] Johnson et al. (1996) examined lake sediments: even in the deepest part, lake sediments are interrupted by a layer of soil containing roots. The lake therefore probably dried up completely, although it cannot be ruled out that small pothole areas may have been present at various times.

adaptive radiation in each of them. Particularly fascinating is the rate of speciation of cichlids in young Lake Victoria.

In addition to the fact that the group is clearly in a plastic phase of evolution, the conspicuous fact that all cichlids spawn in pre-prepared nests and then care for their offspring contributes to this diversity. Such behavior contributes to an elaborate "family care", where members of one form recognize each other in great detail—probably mainly optically—by physical characteristics and ornamentation, as well as by behavior ("dances"). Indirectly, this is suggested by a sad observation of recent years: the lake becomes more and more turbid (due to agriculture on the shores), and fishes must increasingly orient themselves by smell when searching for a mate, but this type of orientation is no longer species-selective, and promiscuous interspecific hybridization appears more and more frequently (Barlow, 2000). Another scourge is the many introduced fish species from elsewhere; these have become invasive and are destroying the native diversity of the lakes.

Sequencing of the genomes of different cichlids naturally finds a number of subtle differences between these presumed species; however, larger differences exist in the regulatory regions flanking the transcribed genes, and in the vast array of regulatory RNAs.

Adaptive Radiation of Lizards of the Genus ***Anolis*** Let's move to the relatively complicated, confusing landscape of the Caribbean islands, where adaptive radiation of lizards from the mainland to the islands has taken place. The island habitats studied are not too far from the mainland (unlike oceanic chain islands), and they host well-established ecosystems that have been formed over tens millions of years. The object of study is a large genus (140 species) of mostly small lizards of the genus *Anolis* (20 cm long on average, half of which being the tail) and their adaptive radiation from the mainland to the Greater Antilles, and from there to the Lesser Antilles, or sometimes also back to the mainland; processes that have been unfolding for about 40 million years. The different ecoforms from different islands (each presumed to have evolved independently) are very similar, so it is a parallel evolution as in the case of the fish mentioned above. Thus, on the four largest islands of the Greater Antilles, we *repeatedly* find 6 anole ecoforms (e.g., Losos, 2009; Feiner et al., 2021; Kavanaugh, 2020): (1) gigantic forms, up to half a meter long, inhabitants of the tree canopy; (2) small inhabitants of the thin branches of trees; (3) inhabitants of the surface of the trunk; (4) inhabitants of the upper floor of the trunk and crown; (5) inhabitants of the lower part of the trunk and base of the tree; and (6) forms living in grass and shrubs.

Morphological diversification has acquired a high rate relative to its mainland ancestors and retained this rate also in lineages that subsequently recolonized the mainland, so that they diversify more rapidly even after their "return" (Feiner et al., 2021). This points to the amazing speed of evolution of some lineages and has yielded a number of insights. Yet "the evolution of island lizards remains a mystery", as Kathryn Kavanaugh's article cited above says.

Butterflies The tropical rainforest can be divided into several layers according to the intensity of the penetrating light. Suchantke (2025) reported that *completely* *unrelated* butterflies in each layer resemble each other much more than they resemble their actual relatives living in different parts of the world, or even just in neighboring layers of the forest a few meters away. Suchantke develops the idea of an ecological organism, which is meant as a whole inhabited by the forest dwellers, who perform certain *fixed functions* and remain in connection with other members of the system, creating a network of functions and activities of unprecedented complexity. The overriding agent may certainly be natural selection as in the case of anoles; he has in mind, however, rather immutable formative laws, not selection from random variations.

One more explanation for these butterfly differences can be offered. Assume an "adaptive landscape" (in the sense of Waddington, 1975), that such an ecological organism actually represents. Such a landscape doesn't just act with forces but also through *traits* that are perceived by the organisms and allow (or challenge?) them to embark on (opt for) an alternative ontogenetic trajectory. It is quite possible that such trajectories exist and are finite in number within the insect order, and so the pheromones, scents, humidity, etc., of a given habitat, could easily redirect the ontogeny. Repetition of the process may then lead to genetic assimilation (see above Sect. 9.2). Parallel evolution is also at play: Van Belleghem et al. (2023) report that very similar wing ornaments (in species of the genus *Heliconius*) are achieved by different "rewiring" of genetic regulation in the background.

Disasters And there is one more very drastic case, envisaged by Matsuda (1987), which leads to a functional flip of the whole biosphere, a Lotmanian "explosion" indeed (see Chap. 4). The adjective "adaptive" does not even fit this type of radiation very well: these are global catastrophes. Imagine, for example, a supernova explosion near the solar system. The hard radiation will kill most life, and most of the remaining creatures will be a collection of various malformations. Those that can reproduce in spite of this will then unleash large–scale adaptive radiation and establish a new biosphere. Another such catastrophe may be dispersing high amounts of pheromone–mimicking substances—for example, drugs in the waters, all applied at once and to the entire biosphere. This is how some paleontologists interpret the adaptive radiation of the Ediacaran fauna at the end of the Precambrian ice ages.

There has always been something; script, tradition, setting, surroundings. Adaptive radiation means "carving out" a new environment where many of the existing settings no longer apply—a (rapid) transition from a multidimensional (adaptive) landscape umwelt to another one. This may happen by chance in plastic groups with evolutionary potential, or in groups that orient themselves in the new environment and move into the plastic phase (see Chap. 11 for plasticity and elasticity). We have attempted to show that the causes of radiations are also multidimensional and pointing out causality of evolutionary events is often difficult. They may be mutations, the use of memory and experience, or new apomorphies. In particular, however, we again point out that the Innerness and interpretative abilities of the living must play a huge role, too.

9.5 Plasticity, Canalization and Reaction Norm

9.5.1 The Charade with a Script as an Entry

In this section, we will conclude our list of interesting possibilities for the transformation of the Appearance of the species. For a century, the various plastic transformations of species Appearances have stood as a challenge to the so-called *Standard Evolutionary Theory* (SET, i.e., the Neo-Darwinian synthesis), which has struggled to find explanations for them. With the influx of new knowledge processed by modern methods, plasticity has ceased to be a collection of curiosities, and its protagonists have been advancing various versions of *Extended Evolutionary Theory* (EET; e.g., Laland et al., 2015) over the past three decades. The aim of the theory is to get rid of considerations that presuppose the innerness of living beings, i.e., all varieties of "vitalism", "Lamarckism", etc. Life must not be conceived by the scientist as "purposeful and creative" (e.g., Futuyma, 2021, p. 360): SET proponents consider the EET activities suspicious, arguing that one day everything will be explained by the standard model. Let us try to rise beyond such disputes; data—especially isolated data—can have all sorts of interpretations, as we show in many places of this text.

Unlike the adaptive radiation and mutation-selection model, we will ask how many phenotypes a single genotype can produce—without mutation and selection. Consider a hypothetical Arctic plant. It lives in harsh conditions that channel its Appearance ("phenotype") into one single form in which it has survived more or less unchanged "since time immemorial". Let us now transfer our plants to the lowlands of the temperate zone, to different soils, different light and temperature regimes, other ecological relationships, etc. The constraints became suddenly released, and the form of the species dances in all sorts of directions, and if we did not know the origin of these variants, we might, after a few generations, confuse them with different species, or even describe them as new ones. In other words, the members of the species seem to be wandering in their umwelt (Markoš & Švorcová, 2019), discovering what is allowed under the new conditions. Needless to say, these are not simple returns to the past, which is no longer with us: the contents of the "memory repository" (the "genetic text") are searched and assessed (interpreted) with the "eye" not of an ancient ancestor, but of contemporary plants shaped for generations by the Arctic. We don't even know if the forms have danced this way due to the stress experienced in the new conditions, or, on the contrary, because they suddenly found themselves in conditions exceedingly favorable, have plenty of resources and therefore don't have to guard their bare survival. These are not mutants: replanting into the Arctic landscape will soon "channel" the plant's form back into a single phenotype. The set of phenotypes that we discover after transplanting to a new environment is called the *reaction norm* and expresses the plasticity of the species, the number of *discrete* phenotypes produced over *a single* genome. Of course, we do not know how many such phenotypes are possible in total. To make this less clear, let us define *plasticity*, a continuous measure of variability around discrete manifestations of a reaction norm.

We are unlikely to find such a "pure" model. After transfer to a new environment, natural selection will certainly come into play and will prefer (in a given habitat) one or the other phenotype. Evolution will involve adaptive radiation, genetic assimilation of plasticity, many mutations, etc.: we just want to show that it is not easy to recognize a really clear "mechanism" behind, for example, the diversification of cichlids in Lake Victoria. Nor would it be easy if we had the genomes of all those forms: we don't know exactly what the relationship is between mutations and the overall Appearance.

Plasticity and the norm of reaction were already studied in the early twentieth century, developmental biology and papers from that time abound with anecdotal accounts with variable terminology, so today's interpretation of observations is quite difficult. Further developments in biology then followed the path that resulted in SET, and the field has only recently experienced new developments. The literature is large, and here we focus on some themes from a set of review articles by authors who have worked in this field over the last 30 years or so (see volume Pfennig, 2021b).

One explanation for the high diversity (wandering throughout the umwelt) caused by the new environment may be the desire to "buy time", i.e. to survive until a more stable (genetic or behavioral) solution to the problem is found. This is the view of Diamond and Martin (2021), for example, who apparently do not intend to move out of the SET mindspace when they conclude their paper by noting that attention must be focused on "subsequent processes, namely whether plasticity does indeed mean buying time for evolution to occur, on the way plastic responses (especially maladaptive ones) affect the evolutionary response" (Diamond & Martin, 2021, p. 203). Does it follow that plasticity itself is not an evolutionary process? In a way, the authors are right—after all, most organisms switch their Appearance in ontogeny according to external conditions, and also according to signals both internal (hormones, pheromones, kairomones, quorum sensing, structure, nutritional status, presence and composition of symbionts, temperature, light regime, sound signals, attitudes, etc.) and external (from seasonal influences to the presence of pheromones and other environmental cues). A well-known example is polyphenism (the different appearance of generations) under different environmental conditions, e.g., spring and summer generations of a butterfly, or winged and wingless aphids. Ledón-Rettig and Ragsdale (2021) therefore consider the manifestation of plasticity as the result of a purely physiological response to a cocktail of various environmental stimuli—external and/or internal.

Ontogenesis is usually an irreversible process, but intergenerationally, the induced change can be reversible—in the presence of other stimuli—or "reverberate" over several generations (more on this in Chap. 10 where, among other things, epigenetic inheritance is discussed). Think of the locusts, which switch their ontogeny to flying forms at high population densities and then invade new habitats, only to return to a "sedentary" life in the next generation. Adult females of some fish, if males are absent from the group, can change their sex to serve as offspring producers.

Similarly, toads of the genus *Spea* produce both "normal" tadpoles, feeding on detritus and algae, and predatory ones, preying on plankton; both types of tadpoles differ considerably in appearance. Here, however, the situation is more interesting: one species of toad switches ontogeny according to reigning conditions, while another species (in a different habitat) almost always has only predatory tadpoles, although occasionally it can also adopt a detritophagous pathway. Here, within the genus, we can already talk about evolution in a certain direction.

Thirdly, the well-known example of Mexican salamanders (e.g. genus *Ambystoma*). Salamanders from desert lakes had to face drying up of the land, so that adults gradually lost their fitness. Therefore, the species became paedomorphic, i.e. locked in their development in a stage of sexually mature larvae that remained in the water throughout their lives and did not need to come ashore. Yet in many species the ability to metamorphose to the land form still remains, but the appropriate hormone to trigger it is missing: many species respond to artificial addition of the hormone and metamorphose to the "original adult" stage. There are a number of similar cases where purely physiological regulation leads to unusual forms, and yet this is not evolution; if such phenotypes are capable of reproduction, they may revert to a typical form immediately in the next generation. Are toads or axolotls "buying" time to bridge the time before they transform into purely aquatic creatures and "forget" the "adult" form entirely? Or rather, they do not forget it, they just exclude it from the places normally visited in frames of their umwelt: various accidental atavisms (the tail in humans) or rediscoveries (the wings in many species of stick insects) convince us of this.

The phenotypic plasticity may be completely lost, and due to natural selection, we get an unchanging species, or such inability of changing is only apparent and the plasticity reappears in a new environment, as in our thought experiment with the Arctic plant. And it may also be a completely new type of plasticity, induced ad hoc in a new environment (Scheiner & Levis, 2021; Pfennig, 2021a). West-Eberhard (2021, p. xviii), in her introduction to the above-mentioned volume (Pfennig, 2021b), further distinguishes between the two types of selection process mentioned above: genetic accommodation directs towards a certain phenotype (Appearance) while maintaining plasticity, whereas genetic assimilation rather suppresses plasticity. Since the terminology is not yet settled, most of the authors in the volume do not distinguish and stick to genetic assimilation, although of course they often deviate from Waddington's original formulation. Levis and Pfennig (2021), in their chapter, then speak of "*plasticity-led* evolution". Here we encounter the notion of evolution that is not *primarily* dependent on changes in the genome: "Plasticity-led evolution can be defined formally as the evolutionary process that occurs when a change in the environment triggers a change in phenotype via phenotypic plasticity, and following the uncovering of cryptic heritable variation in how different individuals respond to this environmental change, the environmentally induced phenotype is refined by selection into a new adaptive form" (Levis & Pfennig, 2021, p. 215). This is also emphasized by Schlichting when he points out the absence of selection, especially because we cannot actually know the extent of the response norm in advance: "Within the normal range of environments, novelties in canalized developmental pathways are most likely to be produced via mutation. However, outside of those environments where canalization has evolved, reaction norms are inherently plastic and thus also inherently prone to produce phenotypic novelty" (Schlichting, 2021, p. 368). An example of broad-based approach to the problem of cryptic variation is the work of Zan and Carlborg (2020). The authors examined 4390 yeast clones and grew them in 20 culture media. They looked at 130 loci (sites in the genome containing a number of genes together) that explained most of the inheritance in all environments studied. This cooperation of a large number of genes was reorganized by the environment. The cryptic variations thus revealed by the environment then provide material for further selection.

Thus, we get to the inheritance of new forms and their transgenerational maintenance. Here, an "intermediate step" emerges in the form of epigenetic regulations and their transgenerational inheritance, which may be a bridge linking plasticity and reaction norm to the classical model of genetic regulation.

References

An, N. A., et al. (2023). De novo genes with an lncRNA origin encode unique human brain developmental functionality. *Nature Ecology & Evolution, 7*, 264–278.

Ankersmit, F. R. (1994). Six theses on narrativist philosophy of history. In *History and tropology. The rise and fall of metaphor* (pp. 33–43). University of California Press.

Barfield, O. (1973). *[1928] Poetic diction. A study of meaning*. Wesleyan University Press.

Barlow, G. W. (2000). *The cichlid fishes*. Perseus Books.

Berlin, B., & Kay, P. (1969). *Basic color terms. Their universality and evolution*. University California Press.

Burda, H., Bayer, P., & Zrzavý, J. (2014). *Humanbiologie [Human biology]*. Eugen Ulmer.

Crick, F. H. C. (1958). On protein synthesis. *Symposium Society for Experimental Biology, 12*, 138–163.

Dawkins, R. (1976). *The selfish gene*. Oxford University Press.

den Hoed, J., et al. (2021). Molecular networks of the FOXP2 transcription factor in the brain. *EMBO reports, 22*, e52803.

Diamond, S. E., & Martin, R. A. (2021). Buying Time: Plasticity and Population Persistence. In D. W. Pfennig (Ed.), *Phenotypic plasticity and evolution. Causes, consequences, controversies* (pp. 185–209). CRC Press.

Feiner, N., et al. (2021). Evolution of the locomotor skeleton in *Anolis* lizards reflects the interplay between ecological opportunity and phylogenetic inertia. *Nature Communication, 12*, 1525.

Fisher, S. E. (2019). Human genetics: The evolving story of *FOXP2*. *Current Biology, 29*, R65–R67.

Franzen, A. (1969). *A concise history of the Church*. Herder & Herder.

Futuyma, D. J. (2021). How does phenotypic plasticity fit into evolutionary theory? In D. W. Pfennig (Ed.), *Phenotypic plasticity and evolution. Causes, consequences, controversies* (pp. 349–366). CRC Press.

Hayden, E. C. (2009). The other strand. *Nature, 457*, 776–779.

Johnson, T. C., et al. (1996). Late pleistocene desiccation of Lake Victoria and rapid evolution of Cichlid fishes. *Science, 273*, 1091–1093.

Kavanaugh, K. D. (2020). Evolution of island lizards remains a mystery. *eLife, 9*, e62230.

Kay, P., & Maffi, L. (2000). Color appearance and the emergence and evolution of basic color lexicon. *American Anthropologist, 10*, 743–760.

Kay, P., & McDaniel, C. (1978). The linguistic significance of the meaning of basic color terms. *Language, 54*, 610–646.

Koselleck, R. (2004). *[1979] Futures past: on the semantics of historical time*. Columbia University Press.

Laland, K. N., et al. (2015). The extended evolutionary synthesis: its structure, assumptions and predictions. *Proceedings of Royal Society B, 282*, 20151019.

Ledón-Rettig, C. C., & Ragsdale, E. J. (2021). Physiological mechanisms and the evolution of plasticity, In D. W. Pfennig (Ed.), *Phenotypic plasticity and evolution. Causes, consequences, controversies* (pp. 113–137). CRC Press.

Levis, N. A., & Pfennig, D. W. (2021). Innovation and Diversification Via Plasticity–Led Evolution. In D. W. Pfennig (Ed.), *Phenotypic plasticity and evolution. Causes, consequences, controversies* (pp. 211–240). CRC Press.

Losos, J. B. (2009). *Lizards in an evolutionary tree. Ecology and adaptive radiation of anoles*. University of California Press.

Marchese, F. P., et al. (2017). The multidimensional mechanisms of long noncoding RNA function. *Genome Biology, 18*, 206.

Markoš, A., & Švorcová, J. (2019). *Epigenetic processes and the evolution of life*. CRC Press.

Matsuda, R. (1987). *Animal evolution in chaging environments, with special reference to abnormal metamorphosis*. John Wiley & Sons.

Pfennig, D. W. (2021a). Key Questions about Phenotypic Plasticity. In D. W. Pfennig (Ed.), *Phenotypic plasticity and evolution. Causes, consequences, controversies* (pp. 55–88). CRC Press.

Pfennig, D. W. (Ed.). (2021b). *Phenotypic plasticity and evolution. Causes, consequences, controversies*. CRC Press.

Pinson, A., et al. (2022). Human TKTL1 implies greater neurogenesis in frontal neocortex of modern humans than Neanderthals. *Science, 377*, eabl6422.

Reber, A. S., Baluška, F., & Miller, W. M. (2023). *The Sentient Cell: The Cellular Foundations of Consciousness*. Oxford University Press.

Richerson, P. J., & Boyd, R. (2006). *Not by genes alone: how culture transformed human evolution*. University of Chicago Press.

Romanidis, J. S. (1981). *Franks, Romans, Feudalism, and Doctrine*. Holy Cross Orthodox Press.

Rubin, C.-J., et al. (2022). Rapid adaptive radiation of Darwin's finches depends on ancestral genetic modules. *Science Advances, 8*, eabm5982.

Rutherford, S. L., & Lindquist, S. (1998). Hsp90 as a capacitor for morphological evolution. *Nature, 396*, 336–342.

Scheiner, S. M., & Levis, N. A. (2021). The loss of phenotypic plasticity via natural selection: genetic assimilation. In D. W. Pfennig (Ed.), *Phenotypic plasticity and evolution. Causes, consequences, controversies* (pp. 161–181). CRC Press.

Schlichting, C. D. (2021). Plasticity and evolutionary theory: Where we are and where we should be going. In D. W. Pfennig (Ed.), *Phenotypic plasticity and evolution. Causes, consequences, controversies* (pp. 367–394). CRC Press.

Schluter, D. (2000). *The ecology of adaptive radiation*. Oxford University Press.

Sonneborn, T. M. (1950). Heredity, environment, and politics. *Science, 111*, 529–539.

Stedman, H. H., et al. (2004). Myosin gene mutation correlates with anatomical changes in the human lineage. *Nature, 428*, 415–418.

Suchantke, A. (2025). *Insect Forms and Patterns. Exploring the Language of Nature*. Nature Institute.

Tomkins, S. (2005). *A short history of Christianity*. Lion Books.

Van Belleghem, S. M., et al. (2023). High level of novelty under the hood of convergent evolution. *Science, 379*, 1043–1049.

Vašíček, Z. (1994). *L'archéologie, l'histoire, le passé: chapitres sur la présentation, l'épistémologie et l'ontologie du temps*. Kron.

Waddington, C. H. (1975). *The evolution of an evolutionist*. Edinburgh University Press.

West-Eberhard, M. J. (2021). A perspective on plasticity. In D. W. Pfennig (Ed.), *Phenotypic plasticity and evolution. Causes, consequences, controversies* (pp. ix–xxi). CRC Press.

Zan, Y., & Carlborg, Ö. (2020). Dynamic genetic architecture of yeast response to environmental perturbation shed light on origin of cryptic genetic variation. *PLoS Genet, 16*, e1008801.

Chapter 10
Of Seeds and Forms

Abstract Regulation of gene expression is an example how sources of information ("seed") cooperate at the level of structures ("forms") building the specific Appearance of both Closures. Virtual information implemented in DNA is made available through a dynamic structure (e.g., enhancers) comprised of hundreds of specimens of proteins and RNA. Another system of protein "writers, readers, and erasers" attends to further, very subtle settings of this epigenetic regulatory system. The result is precise timing and adjustment of cellular and supracellular processes, structures, and regulations, from cell cycle, cellular structure, embryogenesis, up to nuptial ornaments and flowering. After drawing some analogies from our experience with human spoken and written language, we return, in the last section to communicative and collective memory of the living, with its remembering, amnesia, anamnesis, serendipity, skews…—in short to processes that shape the world of all of us living beings. The role of "horizontal" transfer of information (genetic or else) and introgressions by interbreeding between lineages also plays a role in evolution.

Keywords Communicative and collective memory · Inheritance of acquired characters · Regulation of gene expression · Epigenetic markers · Evodevo

> For since we are the outcome of earlier generations, we are also the outcome of their aberrations, passions and errors, and indeed of their crimes; it is not possible wholly to free oneself from this chain. If we condemn these aberrations and regard ourselves as free of them, his does not alter the fact that we originate in them. The best we can do is to confront our inherited and hereditary nature with our knowledge, and through a new, stern discipline combat our inborn heritage and inplant in ourselves a new habit, a new instinct, a second nature, so that our first nature withers away. It is an attempt to give oneself, as it were *a posteriori*, a past in which one would like to originate in opposition to that in which one did originate:—always a dangerous attempt because it is so hard to know the limit to denial of the past and because second natures are usually weaker than first.
>
> (Nietzsche 1997, p. 76)

We draw inspiration for the title of this Chapter—and of our whole book—from the writing of the fourth century Latin theologian and philosopher Aurelius Augustinus,

A. Markoš, J. Horský, *The Seeds of Forms and the Forms of Seeds*, Biosemiotics 12, https://doi.org/10.1007/978-3-032-16131-4_10

moreover in a translation into contemporary English—and dare to use them in a completely different context. Indeed, we believe that this cluster of words could rank among the most important definitions of life today. The whole chapter could be also entitlet more prosaically "Biosemiotic fundamentals at the molecular level".

Box. Augustinus Aurelius, On the Divine Community 5, 11
"Semina formarum, formae seminum, motus seminum atque formarum."

Translation (after Augustine, 2021):

"The seeds of forms and forms of seeds, and the motion of seeds and forms."

Now, the original reads something like this:

"seminaformarumformaeseminummotusseminumatqueformarum".

The diacritics (spaces between words, capital letters, etc.) became only a later invention of the Carolingian Renaissance of the eighth century; the "something like this" refers to the fact that the original manuscript has not survived, and the sequence of letters may have changed over the centuries, due to copying errors and changes in spelling or grammar. In this (rather simple) case, it is likely that there is little in question with concern to such alternatives, but in the case of other texts it is not so simple. As to translation from Latin: we stick here to canonically translated "forms" though the term "Appearances", as introduced in this book, would sound in our ears better.

10.1 Seeds and Forms

In this world, copiable and storable information must be implemented in a carrier, and the carrier is of a "material" nature. Its structure must be adapted for the implementation of information, and also for recalling it when necessary. Below, DNA will be discussed as an example of such the carrier ("seed") of information.

The shape of the double helix of DNA is not perfectly symmetrical, its surface ("envelope") is somewhat "bumpy", depending on the order of base pairs: each pair has a different outline or shape which is, moreover, also influenced by the context of its neighbors in the string. These surface irregularities are specific enough to be recognized by (also specific) regulatory proteins and/or regulatory RNAs: each such bump thus represents a ligand (several base pairs long) to which a regulator can bind. Thus, a DNA molecule is wrapped with regulators ("readers" —see below) that decide in what context a given stretch of DNA will be accessible for transcription. Let's illustrate such a stretch schematically:

...xxExx...xxExxexxExxExxExxxxP<u>G</u>TxxExx... **xxExx**...

Here, the ***G*** represents the *gene* itself, i.e. the transcribed region (usually hundreds or some low thousands of bases long), flanked by the *promoter* region **P** where the protein regulators plus RNA polymerase may become attached; and **T,** the *terminator*, the site at which the polymerase stops and disconnects from the DNA, together with the newly synthesized RNA. The **E**s are the so-called *enhancers*, binding (landing) sites for different regulators (readers): the order of bases in them may differ, i.e., they may bind a different type of regulator each.

Regulatory motives can occur also inside the **G** segment itself (i.e., the segment serving two different contexts), and often also thousands of bases away, or even on another string of DNA. The regulatory molecule may be able to bind more motives on different location, forming bridges that bring together many distant parts of the string. Through enhancers, genes are "pulled" into contexts of a large number of proteins (and genes), forming a dynamic 3D structure of *chromatin*, a complicated tangled web of DNA, proteins, RNA, and a plethora of small molecules. The linear sequence of letters above is but a very imperfect scheme.

The constellation of enhancers and the promoter determine whether the RNA polymerase is allowed to start transcription (to switch it "on"), and at what speed it should work. Everywhere there is a play of molecular shapes: the actual "seeds of forms", The pure, transcendental information stored in DNA medium as the sequence of characters, becomes part of the nonrandom information medium of a different nature. Forms (Appearances) begin to emerge already on this level.

Two reminders before we proceed to the next phase of form "shaping". First, enhancers can mutate (change their form), and thus their accessibility for regulators can change, or they can bind a different regulator due to such mutation—and all of this can have the consequences of changing higher-order contexts.

Box. The Interplay of Forms at the "Lowest" Level

Example one (we follow the review by Snetkova et al., 2022). There are about 500 so-called *ultraconservative* regulatory elements on the DNA strings in humans, which do not differ in base order from those in the mouse or rat and differ only minimally in other vertebrate classes; similar deeply conserved elements can be found also in other large groups of organisms. They were discovered at the beginning of our century, but to this day there is no unanimous interpretation of their function (and the reason of such a strict conservation). They are apparently protected from mutation (natural selection does not allow their variation), and most of then are not present in protein-coding regions. They probably contain a manifold of enhancers (so far confirmed in more than half of elements described) and may play a role, e.g., in some special stages of ontogeny, as switches in life cycles and/or external stimuli. As discussed, mutants are very rare but surprisingly, if introduced by experimenters, they do not threaten life or development (under laboratory conditions). Several such elements were even completely removed from the genome without any phenotypic effect. Hence, these elements probably do not represent some major, deep top-rank codes: other regulatory regions can obviously stand in and serve ontogeny as well. One explanation for their function might be that major reversals in ontogeny need specific constellations of enhancers and regulators, and they tend, and should, be very rapid. Several such regulators could already be "roosting" over a long stretch of the element, sitting in place and waiting for the set to be completed. Thus, they would represent a kind of "hub" that triggers successive phases of embryogenesis and organogenesis. Seeds of forms.

The second example rather suggests an opposite model—that of "forms of seeds". These represent regulatory codes on DNA (such as enhancers) *plus* their regulatory protein(s) and RNAs. Between very distant groups (e.g., sea sponge and fish) we find no correspondence in the "texts" of the enhancers, yet the regulatory proteins (for which we *do* find homologies) trigger a similar sequence of cellular responses in cells from different lineages of organisms (Harmston, 2020; Wong et al., 2020). The obvious point is that the form of an enhancer needs not be determined by a specific base sequence, that an identical or similar *shape* can be reached from different strings of the script, and so regulators land onto different stretches of DNA but of the same shape (as if they "understand" the same text written in different "spellings"). We are here confronted by an ancient regulatory syntax that "means" the same structure: in each lineage, the enhancer–regulator pair must constantly "match" each other so as to regulate the same desired function.

It should be added that protein forms can *also* be obtained from different starting primary amino acid sequences. The whole astronomical number of different combinations of amino acids eventually collapses into about a thousand basic shapes (Dorit et al., 1990); some researchers even posit a kind of discrete universe of protein forms, analogous to the Mendeleev table (Denton et al., 2002; Ahnert et al., 2015). It should also be noted that copies of enhancers (or their slightly different paralogs) can move horizontally across the genome, drawing genes and gene complexes into entirely new contexts. This kind of genomic change, as we shall see, probably plays an important role in evolution, creating new, fruitful contexts. It is possible to adjust such shapes through manipulation of the overall context of the organization of the cellular environment (see, e.g. heat shock proteins, see Sect. 9.2). Below we give two examples of such formation: (1) chromatin structure, and (2) the perception, amplification and interpretation of signals and signs. Their whirling about gives us a clue also to understand other structures and functions in the cell interior.

Chromatin The human genome contains DNA text about three billion bases long, which makes two meters divided into 46 "chapters" —chromosomes; two copies of these (each from one of parents) are necessarily packed into the nucleus of a cell that is 1 μm in diameter. The packing is not random: it must be neat, the individual parts must be accessible (or not) to regulatory molecules, and the regulators must find their binding site in real time: hundreds of proteins and RNAs involved must find each other and associate in functional contexts according to the requirements of the moment (these goals are not an easy task in a crowded and highly structured space). Moreover, when the nucleus divides, the whole apparatus is loosened and re-arranged again in the daughter nuclei—either in the original arrangement or in a new one, thus allowing differentiation. It is important to realize that we don't inherit naked DNA, but such information structures (will we call them "seeds", or do they represent "forms"?).

The first stage of DNA packing into chromatin is the nucleosome, a repeating structure that contains eight *histone* protein molecules in its core, around which a DNA stretch of about 200 bases is wrapped; a short section without histone binding is followed by another nucleosome; the arrangement of nucleosome "beads" does not have a stable position relative to the whole DNA string—this or that section becomes part of one, depending on the cell type as well as and perhaps also on the current demands on its function. The packing continues with other structures, but we will not pay attention to them here.

The basic structure understandably introduced herebut in a very simplified form roughly determines which parts of the DNA will be exposed and thus become "visible" to protein regulators. Histones, again, provide a landing for many other proteins, so that the whole nucleosome section takes a unique shape containing hundreds of different proteins. On the one hand, this serves the gene expression on a given section, on the other it serves the context—the interconnection with other parts of the chromatin even those located at a considerable distance on the linear DNA, perhaps even on a different chromosome.

Signal Particles Let's focus now on the whole eukaryotic cell; for simplicity, picture a round one, about 10 μm in diameter. It has several hundred different proteins—*receptors*—in its plasma membrane that register and filter specific environmental stimuli (hormones, pheromones, nutrients, physical parameters, etc.). Such a stimulus causes a change in receptor conformation, and as a result a cascade of amplifying events is triggered in the cytoplasm. Again, this involves dozens of proteins (an analogy of multiprotein complex is described above for chromatin), and often also the production of low molecular weight substances (so-called second messengers) which are responsible for further amplifying the signal and distributing it to different target(s) within the cell. These targets include the locomotor apparatus, metabolism, chromatin configuration, overall cell shape and position, etc. During a short period of time when the receptor is activated, the recalled proteins are brought together to form a *signal particle*, which can thus link the receptor to its target. Because of the number of simultaneously activated receptors on the membrane, such particles will cause fruitful interference, i.e., they will amplify or quench each other's signals, and they will do so differently in different parts of the cell and at different subsequent intervals. Thus, we get the idea of a dynamic time-shape of the cell, a network that, *as a whole* (along with other structures), registers the here and now of it, and responds to that whole—by focusing on the "adjacent possible" (on the event horizon) towards which it is advisable to move.

Let us add that our examples of multiprotein complexes (chromatin and signaling moieties) are only one of many that can be recognized and studied in the cell. Also, of course, this not only applies to eukaryotic cells; the principle is the same in bacteria and archaea, with their own complexes.

Below we will attempt to shed light on the "motion of seeds and of forms", epigenetic regulations that can project everyday (communicative) experience into the more evolutionarily stable collective memory of the lineage. This will bring a shift of our attention to Closure 2, to cellular or supracellular negotiation (see Sect. 10.5).

10.2 Interplay of Seeds and Forms

Think of texts in a certain language: basic vocabulary, grammar, sentence structure in scientific and legal texts, political speeches, poetry, or fiction. Even when the reader is not an expert in any of this, if she can speak that language, she will understand at least at the basic level any such text, even though the expression (shape, style) and content of these texts differ, and she may not be a fan of poetry nor educated in science or law, neither a bookworm of any sort. Readers will understand them to some specific degree, and this degree will orchestrate the final shape of the work in their minds, even though the "seed", i.e., the text itself, remains unchanged for all recipients. The life history of the individual, its community (morals), or immediate situation(s) will decide its final behavior.

Let us now shift our focus to macromolecular complexes and try to think of them as analogous to literary creations that enable the extraction of meaning. As with written texts, we will not look at these as the merely one-dimensional chains of discrete symbols, but will be interested in questions of the "morphology" of such higher order complexes. The same expressive "language" is thus "understood" differently each time as happens, for example, in the liver, the brain, the muscles..., moreover, with "dialects" (e.g., the liver of a human, of a horse, of a snail...). Such protein complexes co-create the cell, and the cell then acts as a subject endowed with an innerness. We will then discuss the complexes of cells that make up Closure 2—an entity of a different kind than that of the cell, because its "quasi–character" can have different durations, because it can be formed *ad hoc*, and because a single entity can be part of different such quasi-entities aka Closures 2, as for example a multicellular body, a member of parliament, a meadow, or the Czech nation.

The evolution of biological communities (e.g., species or biomes) suggests that the means of gene expression and the understandings of the result can be seen as analogous to the divergence of languages (also of the structured space of understanding, but we will not discuss that here). Related dialects (provincial, professional, etc.) as well as novel languages emerge in which the *same* words (or root words) enter a somewhat different contextual field and begin to appear in different roles; moreover, this is often accompanied by words adopted from elsewhere by so-called "horizontal transfer".

What reading is correct, then? In different languages and in different places in the text, for example, we can find many letters with diacritics (e.g., 'a' as á, à, ä, å, or 'c' as č, ć, ĉ, ç); add to this the differences between lower and upper case letters, and other diacritics such as the space, the full stop, the exclamation mark, the Saxon Genitive, and even more extensive modifications (e.g. ©, $, @).

Thanks to the ability to "get by", for example, an English speaker *somehow usually* understands the speech and texts in other German, and even Latin languages—even though familiar words may mean something completely different in those languages, and the grammar (also in written texts, spelling) is also different. *Somehow,* we can *usually* understand texts from other times (see Chaps. 4–6), e.g., by Shakespeare or Chaucer, or old chronicles. We can see this in our (hopefully successful) use of the archaic meaning of "closure" as an "enclosure" and not the more modern usage of going out of business (of a factory or shop).

We usually understand even garbled texts, though it is not rare to see significant shifts in meaning due to changed punctuation, diacritics, speech melody, etc. We can also manipulate the text in many ways, moving whole sentences or blocks (*transposition*) or inserting quotations from other texts (*horizontal transfer*). In all such cases, the context can be changed, often significantly. But what if an *unusual* understanding also comes to existence, which opens new spaces of possibilities in which the organism (or even line) adaptively moves—or it even gives rise to a new line? In human languages, the poetic license often successfully opens such possibilities (Barfield, 2013 provides beautiful illustrations), and in doing so the poet is merely continuing what life figured out immediately after its inception. *This* is the cosmic event, the great explosion made possible by the existence of closures. Such a fortunate event, a novelty (apomorphy), a stroke of fortune (serendipity), often signifies an *explosion* to new realms that begin to emerge on the horizon of possibility.

The means of the text are further enhanced by the introduction of grammatical norms, which are observed, more or less, in the community, and by expressiveness—both in the choice of appropriate words and (when speaking) in "body language". Leaving body language and vocal capabilities aside, we will draw attention to the possibilities of diacritics in "written" texts, which—while preserving the basic "alphabet" —provide a plethora of subtle changes of meaning and references to pronunciation. Diacritics also include flags marking the beginning and end of words, sentences and paragraphs. To switch to biology: the original possibility of different readings is still maintained, for example in the cells of different tissues, or perhaps according to the season.

Much of what has been said above relates to norms and contexts established within a community, and individuals who may have to struggle to formalize their personal life with the context of community. Let's take for example a lone Indian student in Prague who thinks and dreams in his/her native language, communicates in English with a community of non-Indian (and non-English) colleagues at school, and only rarely (and in a broken way) speaks Czech. About the country and its countryside and history he knows next to nothing. (He is like, say, a redwood tree in a European park, or a tiger in Prague Zoo.) As many times in the preceding text, a caveat: analogies and metaphors *do not imply* identity, but they provide a useful tool: human individuals and communities are alive, and so their characteristics can be used largely to infer how things work for other forms of life.

10.3 Communicative and Collective

In previous chapters, we have cited passim Koselleck's division of historical time into the space of past experience, *somehow* interpreted in the present, and the horizon of expectation on which a person (a cell? a tree?) or community finds a point, and an azimuth to it. In Chap. 3, we then discussed Vansina and Assmann's division of the past space of experience into three discrete and fundamentally different memory sections: The "extant" period (for people about the last 40–80 years) represents *communicative memory* of currently living individuals and their immediate ancestors. "Extant" also involves the plastic, epigenetic adjustment of a population of *cells* and cell communities to the conditions that prevail during the time period to which the epigenetic memory of the cells or community extends. With *this* inertia, entire ecosystems—and perhaps even entire countries—advance into the future (searching the right trajectory towards the horizon).

A strange period of *amnesia*, almost total forgetfulness, sets in regarding the "gap" of almost total oblivion concerning somewhat older periods. People in the community no longer talk about what happened, say, before World War II, and if they know anything about it, it is only from school, from educational programs and literature, or by studying sources. Biospheric units shed this burden in a similar way—their behavior is built around more or less recent or present-day "attractors". The ocean, for example, had to experience a major ecological shock caused by the sudden culling of hundreds of thousands of large cetaceans in the first half of the nineteenth century, but it has clearly found a new modus vivendi since then. Similar examples on the mainland include the establishment of field monocultures on one third of the continents, as well as of vast human settlements.

The communicative amnesia, a "gap" leads towards the third, deep, or "founding" layer of memory, namely *collective memory*, whose temporal boundaries are only approximate and include founding myths and tradition. In evolution, this could correspond, for example, to the basic body plans of multicellular taxa. Only a tiny fraction of the recent—the communicative period—has a chance to become part of the collective one; on the contrary, the rich pool of collective memory allows for the emergence of those parts of it that somehow speak to the extant time; of course, such emergent sets will be altered through interpretation and modification by the contemporaries (see especially Chaps. 5–7). Thanks to the interplay of these two memories (across the gap), it then becomes possible—especially in the turbulent times of extinctions, epidemics, and revolutionary explosions— to define points on the horizon of expectation, and to plot a path towards it. Such are the nationalist, racial or religious outbursts we often witness around the world, for example, but history also offers more fundamental upheavals. Perhaps the most interesting part of what is going on at *present* is the constant confrontation between two kinds of memory: there are always—in the history of life and in human history—conservatives and fundamentalists on the one hand, and progressives and rebels on the other, and both interpret the past and continuation to the future in their own way. And out of all this arises culture, with its stases as well as explosions.

Can we take inspiration from similar divisions of historical time in the field of evolutionary research, or will we be left with superficial anecdotal similarities? We think we can, and in somewhat different ways in each of the two closures—and perhaps we can also identify those fragments of communicative memory that can slip into collective memory. Passively—by concessions to natural selection, and actively—by genetic or other assimilations of various experiences.

Let us consider a slogan "The present is the key to the past", or its biblical version "What has been will be again, what has been done will be done again; there is nothing new under the sun." (Eccl 1,9). Charles Lyell made it a fundamental postulate of his doctrine in his *Principles of Geology* (1830–1833), and thanks to the popularity of the book, it influenced the thinking of scientists throughout the nineteenth century and often to this day. To sum up: geological epochs must be merely a lawful cyclical alternation of well-understood periods. Catastrophes and sudden reversals, according to the proponents of this view, cannot *in principle* have any place in them. The young Darwin was mightily influenced by Lyell—this likely lies behind his position concerning the importance of cumulative but infinitesimal changes in the line of organisms (sudden events not allowed). The slogan "The present is a key to the future" is also frequent in situations wherein certain decisions concerning future development will set a trajectory from which there is no escape for many subsequent cycles. In evolutionary works, the metaphor of the "lord of the mountain" is often used for such situations. The "lord" may not have the highest *fitness* but remains enthroned at the top and will not let anyone else near him.

A bizarre, often cited example of such a situation is the innervation of the neck in giraffes. The *vagus* nerve innervates, among other body parts, the neck and viscera, and the layout of innervation was established in the evolutionary past, at a time when vertebrate body parts (including anatomy) were composed of approximately equal, repetitive segments. In adult mammals, anatomically recognizable segmentation has almost been lost, although it is clearly visible in the embryonic stage before organogenesis begins, and this original setting remains remembered throughout all life. For example, the innervation of the emerging organs follows the original schedule established hundreds of millions of years ago. Parts of a single segment may *now* be quite far apart, but their innervation keeps to the early "map" and preserves the layout of the original topology. The fork of the vagus innervating the pharynx and the thyroid gland, originally branching off from the respective cervical segment, has *now* moved to the level of the heart. Therefore, in mammals the laryngeal branch of the nerve hives off there and must return to the larynx. In most mammals this is not so obvious, but in giraffes this nerve loops an impressive 5 m. It would be sufficient to "make" a shortcut to the thyroid gland just behind the head, but historically it is what it is, apparently it does not matter and so selection does not interfere.

It will remain this way until the giraffes reach their extinction. This is an example *pars pro toto* for frozen, "hardwired" processes in general.

Raup (1991) provides a similar example with trilobites. This species-rich group persisted in the biosphere for 325 million years, with thousands of species, often abundant (they represent three-quarters of the preserved Cambrian fossils). Barring some global catastrophe, one would "normally" expect such a group to have a very low probability of complete extinction. And yet, no such catastrophe is known to have occurred, and so Raup speculatively leans towards the theory of "bad genes". The argument is that the "hardwired" setup (in trilobite ancestors) was perfectly adequate in the Cambrian and for the conditions at the time, but it failed more distant progeny. However, what kind of change might have been involved, and would we recognize and understand it in advance? Consider a thought experiment: all cichlid fishes have characteristics set at the beginning of the lineage, and this standard is easily recognized by an ichthyologist as well as an aquarist. However, it is difficult to predict whether and where the lineage has a similar Achilles heel, which might lead to the extinction of *all* cichlids sometime in the future!

To follow these initial examples, we will again discuss both closures, i.e., the cell closure and the varieties of the inter-cellular closures. We encourage the reader to look for analogies with slogans linking past, present and future, and to do so through the lens of two kinds of memory—communicative and collective.

10.4 Closure 1 (Cell)

Let us repeat: the cellular closure has never ceased to exist during last about four billions of years and all cells carry within them a tradition ("collective memory") of norms established as early as the origin of life. Besides such deeply rooted norms, apomorphies may have emerged in different lineages: "building blocks" of, say, metabolism or structures (see, e.g., Markoš & Švorcová, 2019 in more details). Such novelties were, however, never so drastic that the continuity of either notation ("seeds") or structure ("shapes") could not be recognized. Moreover—and again, since time immemorial there have existed many ways to communicate from cell to cell, either by cell fusion or by established protocols transmitted through the external environment. Thanks to such "horizontal" procedures (we will review some of them below), many of the specific properties of cell lineages are blurred. Analogies with natural language may help to understand the point.

Dictionary A dictionary (a collection, entries of words used at a certain time) is a mixture of very old words, plus neologisms that appear in communicative memory but soon drop out again; only few of these move into collective memory. A "gene" is usually understood to mean a protein-coding *entry* to an analogous "dictionary", to a pool that a species has here and now. It can be even defined as an information contained in a stretch of DNA, say, longer than 100 amino acids that contains the recipe for synthesizing a protein. As explained above, such a gene is recognizable in a long string by the labels marking its beginning and end. It is reported that the average archaeal or bacterial cell contains (in one set) genes defined in this way in multiples of (lower) thousands; in a eukaryotic cell in makes roughly 20,000. We have already mentioned that when applying this method of counting, the human genome is rather a desert in which the oases of genes make up only about 4% of the length of the chain; the rest are various tags, enhancers, structural elements etc., that enable regulatory proteins to organize the "text" into a chromatine structure, to localize the gene and regulate its transcription.

However, the situation is not so simple; a "gene" can *also* mean parts of the genome that are, or can be, transcribed into RNA, regardless of whether they are intended for translation into a protein or not; in such a perspective the greatest part of the genome is transcribed under various occasions, giving rise to a variety of types and functions of hundreds of thousands RNAs. These stretches are not as easily recognizable on the genome sequence because they are not as sharply delineated as "proper" protein-coding genes. The quantitative aspect (as in a language dictionary) of all such processes also plays a role: alleles of the same gene may differ in their transcription rates. Note further that the "dictionary notation" only vaguely refers to "morphology": the resulting shapes of proteins are dynamic and their functioning context-dependent (e.g., Markoš, 2002 for details).

Grammar First, consider the different readings of genes and RNA: a single gene can give rise to different-length transcripts (i.e., RNAs) and thus different proteins. The transcript itself can often be edited in many ways, leading to different entities (of RNA and/or proteins). Finally, the resulting protein is also subject to a number of modifications, including the trimming or attaching of different molecular units, or chemical modifications of amino acid moieties. As a result, the well-known cliché "one gene, one protein" has quite a limited validity, and the set of different RNAs and/or proteins may be orders of magnitude larger than that of genes.

Furthermore, there are frequent changes in the "word sequence", when parts of DNA jump to another place within the genome and cause a change in the phenotype of the cells or supra-cellular assemblages (e.g., Carroll, 2006). Evolutionary biologists also attach great importance to jumps of regulatory stretches in the genome to the vicinity of other genes, which places such a gene in different contexts of time and space (because the gene becomes prone to other kinds of regulation). Such manipulations often drastically affect the evolution of Appearance in a given lineage. In other cases, directed mutagenesis of certain parts of the genome is allowed in some cells: this is how billions of *new genes* encoding immune system proteins are created in humans during ontogeny (compare with the 20,000 we inherit from our parents).

Similarly, the vocabulary and grammar (and thus the Appearance of the cell) can also be influenced by so-called *horizontal* gene *transfer*, i.e., acquisitions from elsewhere that have become part of the genome and have been incorporated into a new context for which they were not originally "intended". Horizontal transfer of genetic information is an essential component especially of prokaryotic life—both bacteria and archaea; so much so that there is often talk of a global genome (*pangenome*) shared by all. In this way a population of prokaryotes can acquire antibiotic resistance, pathogenicity, or the ability to process a metabolite. We find "foreign" proteins in eukaryotic organisms as well: for example, proteins of viral origin function in the placental constitution of both mammals and sharks (e.g., Chuong, 2013). Similarly, legumes have acquired hemoglobin from animals, though here it works in a context of nitrogen fixation, i.e., in a completely different context than in animals.

Box. Horizontal Acquisitions

A case study on insect genomes (Li et al., 2022) will help shed light on the extent of such transfers. The authors compared 218 genomes and identified 741 events that resulted in the horizontal transfer of 1410 genes derived from non-animal groups—mainly from bacteria, fungi or viruses. In butterflies, for example, such horizontal transfers are common, and some are very old, so old that the "foreign" gene became established in most species. The authors identified one of these and determined its origin from a bacterium in the genus *Listeria*. By homological comparison in databases, they show that the gene–encoded protein is homologous to an alcohol dehydrogenase protein (however, the protein itself has not been studied, and what function it serves in the cell—bacterial or of butterfly—is unknown). Authors then took a single species of butterfly and removed this gene from its genome. It turned out (by ethological studies) that such butterflies are clumsy in courtship, which results in a smaller number of eggs, on top of that mostly unfertilized or defective. How all this relates to the function of a given protein is certainly a subject of intense study.[1]

[1] May they be drunk when lacking the alcohol–degrading protein?

Another example of a similar gene intrusion—even older one: One of the key proteins enabling eye function in vertebrates apparently "jumped" some half a billion years ago to their common ancestor—from bacteria (Kalluraya et al., 2023). Other examples of such gene borrowing are given in the main text.

Analogies from language adoptions often suggest similar changes of function. Compare, for example, the name of the Restoration era London cabaret Vauxhall, which has been transferred into Russian (*vokzal*) to mean—railway station. Similar reversals are common in language adoptions when transferred to a new context. For a comparison of this development in languages and prokaryotic life forms, see List et al. (2013).

Interspecies gene transfer between eukaryotes can take place also through the sexual process of *hybridization*—between species within a genus, family, or even a or higher taxon. The hybrid carries genetic traits from both species and—if fertile—may produce offspring with a partner of parent species; in this way, *introgression* of foreign genes or entire groups of genes from another species can occur. *Hybridization chains* (or even entire hybridization *networks* involving many such "promiscuous" species) are particularly common in plants (an estimated 7–10% of all plant species have arisen in this way), but they are not an exception in animals. In plants, genome duplication (or even richer multiplication) is also frequent, thus enabling the creation—out of the hybrid—of a new species that can exist independently, without a necessity of backcrossing with one of the parent species. Interspecific hybrids can even arise by grafting (Fuentes et al., 2014).

Interbreeding that produces a fertile progeny often occurs in zoos (e.g., tiger—lioness or vice versa), but it can also occur in nature (e.g., polar bear—grizzly). It has already been mentioned that in the past our species *Homo sapiens* experienced genetic introgressions from both Neanderthals and Denisovans and has maintained some of them to this day; similarly, other species can interbreed at the line of contact, and a number of similar events have been described in particular for flying animals like birds and insects, or in invasive species like in the genus *Aster* (summarized, e.g., by Pennisi, 2016; Papale et al., 2020).

Edelman et al. (2019; also, Dasmahapatra et al., 2022) identified a number of introgressions acquired over the past five million years or so on two dozen genomes in the butterfly genus *Heliconius*—ones that probably led to major rearrangements in wing ornamentation. Due to such introgressions, it is very difficult to construct (based on genetic data) a phylogenetic tree of the species relationships of the whole genus, and only genealogies based specifically on highly conserved genes show any reliability—otherwise the evolutionary history of the genus may often be quite blurred.

Dialects The simplest example of a "dialect" expressed by a phenotype is seen in the alleles of genes in sexual organisms. Depending on which alleles have come together in an individual, as well as on their quality and penetrance, we can often estimate the phenotype from the genotype. Or rather, such alleles modify contexts by way of a series of epistatic interactions—wherein the resulting phenotype is a product of the "bargaining" of a large set of gene products, i.e., a series of protein allelic variants, enhancers plus a "cocktail" of various regulatory RNAs.

In parallel lineages of organisms (for example, related species or higher taxa), homologous proteins—*orthologs*—can be found that are a result of inheritance from the original ancestor of such lineages. Certainly, they have mutated in evolution, and thus contribute to the difference in "wordings" in the "dictionaries" of the lineages in question. Obviously and of even more importance is the frequent duplications or even larger multiplications of the same (as in families, e.g., of TKTL, Hox, or Fox genes mentioned in Sect. 9.3.1). From this point on, a single gene of such a multiple set is sufficient to provide the original function, while other paralogues can freely and harmlessly mutate, even if transitorily losing the function of the resulting protein: "dialectical" versions, called *paralogues*, of the original gene are created, and their products (RNA or proteins) enter new contexts, often with interesting consequences for the lineage (see, e.g., the "mondegreen" charade).

Diacritics The above-mentioned heterogeneity is only a boring or necessary background for the processes that—using the analogy of writing—come into play with the introduction of "chemical diacritics" at the levels of DNA, RNA and proteins (see Fig. 10.1). Let us call such modifications *epimutations*, chemical modifications of the components of macromolecules. Compare with *mutation*—changing the genetic transcript by replacing one base with an another. When there is a mutation at some point in a strand of DNA, for example C to T, this mutation is in principle irreversible, and will propagate to all the offspring of the genome and the cell, and in the case of a gene encoding a protein, it *can* mean an amino acid substitution—forever, in all generations. We have already mentioned sickle cell anemia as an example of such a mutation event.

Let's compare this with "chemical diacritics", for example of altering the A, C, G, T bases in DNA. As we recalled in the previous chapter, the double helix formed by the non-random chain of these four units (bases) is not ideally symmetrical; its shape will be different for different sequences in the stretch. Take, as an example the strand -xxTAGCTCCxx-. Let us also recall that during replication—copying the chain—the new complementary strand is read in reverse order: in our example, then, we get -GGAGCTA-.

Now let us add an enzyme recognizing the AGCT motif on the strand and attaching (in the instances wherein broader contexts will be met) a methyl group to the "C" base—cytosine will be converted to methylcytosine. Thus, a *fifth* base in the DNA molecule appears, let's mark it "Č". Without doubt, such an *epigenetic* modification or *epimutation* (introduction of another unit outside the basic set) will drastically change the shape of the "C" moiety and consequently the shape of the DNA stretch at the given site, and may cause—alone or in combination with other such modifications—a different arrangement (again a change of conformation, i.e., shape) of the context of dozens of regulatory proteins flanking the DNA at that site. In short, AGČT and AGCT represent two *different* "words" with possibly different meanings, and such an "accentuation" will cause the relevant section of genetic text to be read and understood differently.

Importantly, when copying a DNA molecule, the replication mechanism is oblivious to epimutations, so it assigns new complementary units to the copied strand *as if it* were unchanged, "unaccented". However, proteins comparing the two strands are usually also present in the replication complex, and if a *motif* with a "caron", i.e., AGČT, is detected in the parent strand, methylase C—the enzyme can attach the appropriate modification also to the daughter strand. Thus, the epimutation *can* propagate to subsequent generations of cells (and sometimes even to the progeny of multicellular organisms) as if it were a "regular" mutation. Yet it is recognized and can be—even after several generations—removed—unlike most "base letter" mutations. Moreover, there exists not just a single "methylase C—demethylase Č" pair of such a writer–eraser set: there is a large number of them, for different motifs (i.e., shapes of the respective DNA stretch) in the chain and for different external and internal situations (contexts).

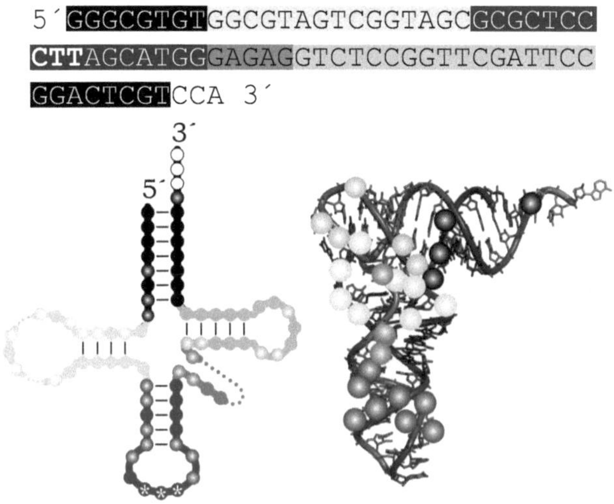

Fig. 10.1 *General scheme of tRNA with some modifications (epimutations) of its sequence.* Above, the primary sequence of one of several hundred genes for different tRNAs; future sections of the primary transcript are distinguished by coloring (as below). Below, the secondary structure of the native sequence, projected in two dimensions, and a three-dimensional scheme of the same, with locations indicated where epigenetic modifications may occur. The number of nucleotides so altered varies according to context, and individual molecules may vary in "ornament" and hence in function (re: Suzuki, 2021)

Moreover, the methylation of base "C" is not the only possible epigenetic modification, methylation is only one of numerous such "accentuations", and other bases can be modified as well. Even richer are the possibilities of such accentuation in the case of RNA, proteins and sugars (see below and Fig. 10.1 for one class of short RNAs), and the set of twenty amino acids in a protein can be multiplied *many times* in this way: the protein will then be composed not of twenty, but even, say, of 80 different amino acids.

Of course, for epimutation (or its removal) to happen, the appropriate enzyme(s) must be present, and such enzymes themselves may also be subject to fine epigenetic adjustment. The list is not infinite, but even so there may be hundreds or even thousands of similar pairs in a single human cell, or one similar one to it—moreover, these adjust to recognize non-basic shapes. Thus, the methylase C from our example can methylate a motif other than the one mentioned, provided that the molecule at this site, e.g., bends under certain conditions, or changes its shape due to the binding of some regulatory protein or to changes caused by some macromolecule in the vicinity. Thus, there is a large (perhaps the most numerous) set of regulatory proteins in the cell that attaches (writers, W) to the macromolecules and introduces modifications, that reads and differentiates between them (readers, R), and that can remove (erasers, E) such changes.

10.4.1 The Fate of One Class of RNA

In this section, we review a group of non-coding RNAs that play various roles in chromatin formation, translation, immunity, etc. (e.g., Suzuki, 2021). Thousands of such transcripts are massively epigenetically modified, and often a failure to synthesize even one of them can upset the homeostasis of a cell, leading to its death. Figure 10.1 schematically illustrates such possibilities in the case of tRNAs (Suzuki, 2021).

Importantly, not only somatic cells but also eggs and spermatozoa contain a large number of different regulatory RNAs, which can after fertilization influence the environment of the zygote, and consequently the development of the embryo. In this way, some of the parental experience *can be* passed on to the offspring via the gametes and *may* persist for several generations (e.g., Švorcová & Markoš, 2023, for more details). If not reinforced in subsequent generations, they gradually disappear, however propagation across several generations may provide room for genetic assimilation.

This approach to molecular epigenetics will be inferred throughout the rest of the chapter—and we should come to the realization that the *vast majority* of life's processes are epigenetic in nature, from deciding which parts of the genetic library to consult and how to interpret the script, to the behavior of the biosphere. Thus, "genuine" genetic processes are limited, more or less, to the mere storage of the "canonic" DNA, its copying, and the basic machinery of transcription and translation. But these processes also depend on the epigenetic organization of the cell, which, as we said in the introduction, "has never before been dead".

10.4.2 The Cell as a Reader

The cell (as well as multicellular assemblages of different kinds, (see below, 10.5) pins down *meanings* of external and internal contexts, filtered through its own memory and experience. The cell thus is a "reader" recognizing the subtle nuances of internal settings. For example, genes that we needed in the fetal stage may be blocked for the rest of one's life, or may be unblocked later but in the context of quite different activities. In this way, "normal" automatic functioning can also be switched to a different mode in the event of, for example, wound healing, infection, or a change in external conditions (expected like seasons, or unexpected like famine). However, errors, from trivial "mondegreens" to developmental disorders, cancerous growths, or ecosystem collapse, are also part of the game.

What does this amazing game actually mean? It means that *there's no need to have too many genes*. The basic, "dictionary gear" ("genetic records"), plus the norms determining how to use it, are sufficient for a cell—with its memory and experience—to relate meaningfully to the world. Just as the dictionary of any language lists only the basic forms of words, plus the rules for bending them (or even forming them, as in the case of numbers or chemical nomenclature), so the genetic memory of the cell does not exceed the necessary size of the database (or program). A slavish "pidgin" based only on the basic stock is sufficient only in special cases, it is compressed into a minimal vocabulary, and we can force the cell to survive in such a state; but when we stop such a pressure, the "imagination" of the body starts again to develop in the form of an epigenetic game. Playing yet adhering to the frame prescribed by rules is...—can we call it a form of culture? A language? We are simply betrayed by the terminology: we are not used to thinking in this way.

10.4.3 Differentiation

Based on the evaluation of external and/or internal conditions, cells can change their Appearance. For example, some species of bacteria will stop dividing when they are endangered by a lack of food, and instead enter a mode of spore formation: this means a fundamental reorganization of the internal structure and/or functioning of the cell. Other bacteria do not form spores, but also switch to a different less obvious mode: for example, it is often not possible to eradicate a bacterial infection because of the existence of dormant cells, persisters inactive at that time, not responding to external influences (Diamond & Martin, 2021).

Eukaryotic cells differentiate in a similar way: this is most evident in multicellular organisms, where the zygote, spore, or stem cells give rise to entirely different cell types; these then proliferate, switch to yet other cell types, or both at the same time. A special type of change is dedifferentiation—the reverse transformation of specialized cells into stem cells (see below Sect. 10.5). Thus, cells move in their morphogenetic space to other niches and according to the space-time coordinates reigning there, they obey the needs of the momentary coordinates of body's umwelt (or even disobey, as in the case of tumors, etc.). This is the beginning of the complex interplay of Closure 1 with the different embodiments and variants of Closure 2.

Research into the differentiation of animal stem cells in vitro towards the formation of so-called *organoids* (millimeter-sized) has become very fruitful in recent years. In tissue culture, it is possible to "grow", for example, a miniature "stomach" with an internal cavity and layers of appropriate tissues in its wall; similarly, a miniature, histologically differentiated "forebrain" can be prepared (Revah et al., 2022), etc. The structures of the cell surface and its surroundings play a crucial role for proper organoid differentiation (Gjorevski et al., 2022).

We have already mentioned that the overwhelming majority of proteins in a cell (e.g., the Writer—Reader—Eraser systems mentioned above) are devoted not to metabolic but to interpretative functions. And we have mentioned that there are hundreds of sensors on the plasma membrane and in the cytoplasm, which all record the parameters of the external and internal environment. These entries interfere with each other, i.e., they amplify or suppress their action according to the context, and the result of the integration of all these processes is a decision (considering, of course, also different settings of memory and experience of the cell) as to which developmental trajectory to undertake. In this sense, the cell is an autonomous agent (sensu Kauffman, 2000), and one of its main functions is negotiating a common trajectory *especially* with the other autonomous agents with which it can interact.

With differentiation, our focus shifts to the realm of Closure 2. There, some functions recursively repeat the processes known from the cell, but with an important difference: in addition to cellular memory, new kinds of memory appear, based on the experience of the whole community. An important apomorphy is that of human civilizations, which have in added to collective and communicative memory entirely new means of storing and copying memory records. This involves new forms of transmission of tradition, experience and interpretative competence.

10.5 Closure 2 (Cell Community)

While a closure of the first type could never—since the beginning of life—have ceased to exist, closures of the second type are of a different nature: they arise, transform and disappear, with varying importance and differing durations. Despite their mutability and temporality, Closures 2 are *also* characteristic of life from the dawn of life.

First, let's try to think of hypothetical "autistic" cells that live without any "membership" in any Closure 2: while they are well insulated from their environment and control their interior, they only respond to external *physical* factors like temperature, light, salinity, nutrients. This means that while the plasma membrane can filter flows between the outside and the inside, it does not sense the presence of other cells in the environment. This situation is somewhat similar to that of the laboratory prison of bacterial culture described in Chap. 8—but only similar. We don't just mean that such prisoners must be cared for by other inhabitants of the biosphere, humans in this case; without them the laboratory culture could not persist for a longer period. We also point out that even in such a reduced community, cells that previously survived in the biosphere for four billion years carry with them a considerable part of their evolutionary ("collective") memory and experience, and also are aware of each other. In addition, as mentioned in Chap. 8, some types of such communities can set up a division of labor and establish a simple ecosystem. Simply put, autistic cells do not exist in the biosphere, and it is questionable whether they ever did.[1] Let us observe how various conservative norms of a new type—dependent on the cooperation of the community and determining its behavior—emerge in evolution.

Recall that we seek to remain as far as possible from notions of hierarchies ("ranks") of "higher" and "lower" organisms. In our interpretation, every community is first and foremost a member of nested, interconnected *network* of

[1] Except for brief periods when they have been thrust into a uniquely isolative situation—for example, bacteria of a single species swept by water into the depths of the Earth's crust (Chivian et al., 2008).

intercellular collectives, be it a bacterial mat, a religious community, an oriental despotic state, a liver tissue, or a holobiont. The dialectics of "immortal" Closure 1 and time-limited Closures 2 stand in opposition to each other, however analogous their vital manifestations may seem. We ask what *all* supracellular life forms have in common, and our answer is that they weave a story (a plot) based on cognitive and communicative memory i,e,a *culture*. In their way, they *understand* the world and influence it, the world that is only possible because of the continuous existence of Closure 1. They understand the world in the form of a narrative, woven of the communicative and collective memory of the community, of individual narratives shaped by the different types of closure, and also of the sudden "explosions" that can turn that "narrative" into a completely different mode. We seek to glimpse—and hopefully the reader will also see—life from a biosemiotic perspective rather different from that of established, certainly venerable and necessary categories (human—non-human; nature—nurture etc.). This is also the reason for the seemingly illogical cut with which we have defined Closure 1 and Closure 2.

10.5.1 Cohabitation

By symbiosis, of course, we do not mean mere "deposits" of random organisms (as is a Zoo, a crowded bus, a bag of beans or laboratory models), but communities of cells and supracellular units that communicate with each other and thus can meaningfully interact and build their common world. As can be seen, such communication can only occur through shared norms—such that information produced by organisms is *somehow* received and *somehow* evaluated by others. This is also how multi-species mixed communities, often containing dozens of species from all three domains of life, i.e., bacteria, archaea and eukaryotes, communicate. Thus, they must share the ability to produce certain signals, signs or traits, as well as to receive and respond appropriately to them. Some examples are given, though apparently none are present in a pure form; communities use many such means simultaneously.

Quorum sensing is a communicative ability based on pheromones—signaling substances that inform or otherwise manipulate the recipients (location, food, sexual attraction, necessity of elimination of wastes, etc.). These forms of communication are extremely sophisticated and hold many communities together. Another form of communication is vesicles that are pinched off from one cell type and fused with other cells. The contents of such vesicles can include proteins, signaling agents, RNA, and other components of the cytoplasm (Biller et al., 2014; Melentijevic et al., 2017; Cai et al., 2018; Toyofuku et al., 2023); strictly speaking, this is also how eukaryotic viruses spread.

Finally, various forms of *extracellular* matrix—cell walls, gels or networks composed mainly of proteins and polysaccharides (i.e., polymers of sugars), sometimes even minerals—constitute very important components of communities. They undoubtedly function as the support of the body, for example, wood, bone, skin and ligaments, and in the past, they were considered solely as such. However, in addition they are carriers of various structural and epigenetic markers (signs, signals) that surrounding cells can palpate with the help of specific receptors, and regulate their mutual behavior accordingly. They move, grow, differentiate—as well as continuously modify the properties and "messages" of such equivalents of rock paintings and billboards. In particular, the ontogeny of multicellular organisms would be inconceivable without intercellular matter, see below the Box on the neural crest. However, the importance of the extracellular matrix is self–evident in all organisms. For example, multispecies bacterial coatings build their richly structured "dwellings" of extracellular matrix by the concerted effort of all members of the assemble—and react to its structure accordingly.

Coordination of cells in multicellular clusters is also achieved in Closure 2 with the help of other tools, such as (1) positional information, i.e., cells informing their neighbors of their mutual position and differentiation; (2) related and jointly secreted and structured extracellular matrixes, which feed back to the cells their "place"within the body; (3) diffused signaling substances such as hormones, pheromones, and morphogens; and (4) reciprocal rearrangement (especially in animal cells capable of movement) and ingrowth of long cell protrusions into other tissues and organs (e.g., neurons, or fungal mycelia). Cultures? Yes.

Box. Neural Crest and Morphogenesis

In the so–called neurula stage, the vertebrate embryo is segmented into the heart, the digestive tube and the neural tube, which later gives rise to the brain, spinal cord and spinal ganglia. On the dorsal side of the neural tube, the neural crest differentiates, from which cells start to actively migrate: these give rise to the facial part of the skull (bones, muscles, teeth, etc.), and in the body segments they organize the formation of the lower layers of the skin (melanocytes, etc.), and also give rise to some neural ganglia (e.g., Hall, 1999).

The cells of the neural tube have already—during the differentiation of the neural tube (they will not, e.g., place the head in the middle of the trunk) — received information about their position on the antero-posterior axis of the body. As they travel along the sides of the embryo forward and towards the ventral side, they acquire more and more detailed information, and adjust their differentiation in a certain direction accordingly. This information is mainly provided by the "ornamented pavement" of the embryo's surface cells, consiting from a layer of structured polymers, proteins and polysaccharides. Cells passing above the layer read, but can also modify this sugar code, leaving a message for the next sequence of incoming cells (so that the code mosaic varies over time). An idea of this incredible heterogeneity and the WRE system is given by, e.g., Gabius (2018) or Kyprianou et al. (2020).

Below we provide examples of the symbiotic relationships of the Closure 2 type. We mention just a few, leaving aside human ethnicities, religious communities, nations, languages, states, etc. We will discuss four groups: 1. multicellular organisms; 2. close coexistence of two or more species; 3. ecosystems; 4. intracellular symbioses. The list is far from exhaustive, these are only examples, and they are not given in any hypothetical "evolutionary order" —the arrangements described are not "pure" in almost any organism. And once again we stress that this includes human culture based on mind, speech and writing. We have devoted a considerable part of the previous chapters to this, and it stands—as everywhere in this book—in the background of our analogies.

Nothing of what we have described here would be possible without the existence of Closure 1 and its evolution; however, we invite the reader to observe how Closures 2 evolve along lineages and retroactively affect the differentiation and evolution of the cellular interior; we also note the embedding, cooperation, and interdependence of different types of closures, depending on the elicitation of contexts and their confrontation with norms. In each lineage, new types of behavioral norms are established in Closure 2, and these evolve in ways similar to the norms of Closure 1; the difference is that during ontogeny communities of cells or multicellular communities gradually emerge and establish themselves, but also disappear. Closure 1, of course, also differentiates, but it must necessarily ensure its continuity—and therefore the cooperation of both closures must eventually lead to the production of propagulae. The question constantly gnawing in the background, of course, is this: can Closure 2 break into Closure 1 and thus insert certain components of its memory and experience into cellular memory? We think so, but we acknowledge that the sources and clues we work with are not always conclusive (for collective memory in human societies, see Halbwachs, 1980).

Multicellularity The ontogenesis of a multicellular body is an excellent example of the cooperation of the cell interior (Closure 1) with a whole range of Closures 2 established in the interest of the coordination of cells within the body, of differentiated stability and importance; tissue cultures, non-microbial organisms, etc. are interesting models for study. A single cell—a zygote or spore (sometimes an unfertilized egg or other less standard means)—will give rise to the bodies of animals or plants, but bacterial colonies may also arise in this way, and a clump of hyphae will in turn give rise to a fungal fruiting body (Moore, 1998).

Let us first consider origin from a single cell, which is theoretically "responsible" for the Appearance of most aspects of the resulting organism (together with the influence of the inanimate environment). Note a principal peculiarity of this type of development: it seems that the very early stages of multicellular bodies must be carefully isolated for a certain period of time from the influences of the rest of the biosphere, and so various kinds of envelopes appear to ensure the undisturbed autonomous development of early embryos.[2] During this stage, the embryo is

[2] The same applies to multicellular formations of bacteria: colonies. They do not have embryogenesis in the strict sense but can only arise (from a single cell or even a group of cells) in an environment without the presence of other bacteria in the immediate vicinity. In such circumstances, multi-species coatings, consortia, etc., are formed.

dependent on "automatisms" stored in genetic memory and modulated by epigenetic memory acquired from both parents; in some cases, the influence of the mother's organism is also apparent—for example, in the nidation of a mammalian egg in the placenta, or in the embryogenesis of a seed plant embryo in the germ sac. After this protective period has elapsed, the new organism (and this includes fungal fruiting bodies and the bacterial colonies) in turn begins (and must begin) to establish eagerly relationships with other inhabitants of the biosphere, as discussed below. There are many inferred processes of the ontogeny of multicellular organisms—development from cuttings, tissue or organ cultures, etc., we will mention just few of them.

Chimeras of various designs are worthy of attention. They are aggregations of different individuals, sometimes belonging to different species, cooperating to such an extent that ontogeny culminates in a functional body, often also producing functional propagules (but not common sex cells). A paradigmatic example of such chimeras are the *lichens*, a large group of organisms practicing, from an evolutionary point of view, long-term (i.e., not *ad hoc*) coexistence of two or even three unrelated species of organisms (algae, filamentous fungi, probably also yeasts). *Ad hoc* chimeras also include grafted plants, which can also arise in nature by the accretion of trunks or roots.

Chimeras can also arise (even naturally) in mammals: two embryos in utero can fuse in early stages to form a single organism, or only their fetal envelopes with a common bloodstream can fuse, so that the cells of both fetuses can mix. In laboratory experiments, even interspecies or intergeneric chimeras can be constructed by grafting tissues or organs onto fetuses. These experiments point to deeply conserved developmental norms: transplants recognize each other's signals in the chimera partner and construct a viable body: for example, chicken with quail organs or tissues, or even a head (Hall, 1999). Female mammals are an interesting case of chimerism. Their body is a mosaic of body parts containing an active X chromosome derived from either paternal or maternal genomes. If the two chromosomes carry different alleles, this mosaicism can sometimes be reflected in the external appearance: for example, tri-colored kittens are, with rare exceptions, only female. Also, the placenta consists of chimeric tissue from maternal and fetal cells. And a minor quirk to the topic: it turns out that during pregnancy, fetal cells enter the mother's bloodstream through the placenta and settle in various parts of her body (even manipulating her brain); thus, after each subsequent pregnancy, the mother becomes an increasingly complex chimera of her own cells and those of her offspring (Boddy et al., 2015).

The last type of chimeras we will mention are those prepared by experimental medicine. We are not referring to the successes of organ transplantation (of course, these also belong here), but to developmental chimeras, where organoids are implanted into an early embryo and develop therein, obeying the norm of vertebrate development. Thus, for example, a human cerebral cortex organoid can take up residence in the developing mouse brain and form a "humanized" region there, in which the development of the human cerebral cortex can be studied: transplanted neurons establish—in the developing mouse brain—connections with its various parts and build a chimeric brain. The chimera can be used to study the effect of targeted mutations, or in vivo processes that lead, for example, to parkinsonism or dementia (Revah et al., 2022). Similar experiments on the brain of a living human are of course not possible. Another line of research explores the possibility of growing whole human organs for transplantation (for example, whether it is possible to grow a human heart in a pig).

We have already said that cells in a developing organism are sensitive to the context of their environment and adjust their behavior accordingly—from gene switching to metabolism, to signal transmission and the interpretation of signals coming from the outside. In addition, the most differentiated cells (muscle, epithelial, bone, nerve, etc., if we are talking about animals) are not able to "reverse gear", to dedifferentiate, or if they do, only by one or two steps (for example, in wound healing). This is not the rule everywhere: in plants and fungi differentiated cells easily switch to another differentiated type. Also, most differentiated plant cells in can be transformed into a formless tissue culture state (callus); and from such a state they may give rise to embryos and young plants.

The research on embryogenesis in animals has developed in many directions: the individual stages in common animal models such as the fruit fly, nematode, zebrafish, mouse or chicken are known, and thanks to this we know that the basic regulatory circuits are largely homologous, and the subject of evolution becomes one of the interpretation of these tools in the construction of multicellular structures, or epigenetic play over "basic texts", which are often very similar; see below the Evodevo section.

As we mentioned, in animals—unlike plants—dedifferentiation of tissue cells into pluripotent stem cells or even totipotent sex cells is not easy. Nevertheless, undifferentiated stem cells remain in the body to replenish damaged or dead cells, but they usually do not have the opportunity to differentiate into cells of other tissues. This is why so much media attention was given in 1997 to a sheep cloned not from a zygote but from a special construct: they removed the nucleus from an unfertilized sheep egg and inserted in it a nucleus from a somatic (i.e., differentiated, specialized) cell from a mammary gland of another sheep (Wilmut et al., 1997). In the cytoplasmic environment of the egg, the somatic nucleus was able to unblock its developmental restraints, it again became "totipotent", i.e., it set (or reset?) itself to the start, as if it were a nucleus formed by the fusion of the egg and sperm nuclei. This artificial "zygote" then began its normal development "in the test tube". After reaching the blastocyst stage, the embryo was then planted into the uterus of a live sheep, which then bore it.

The second branch of mammalian embryogenesis research does not follow the path of constructing an "artificial zygote", but takes inspiration from tissue culture and organoids, and tries to prepare model embryos (embryoids) *ex utero*, and *then* study them to a high stage of development. Since there is no need for fertilization and the (quasi-)embryo is constructed from tissue culture, many ethical issues are removed, so that it is possible to work even with constructs grown from human cells grown in tissue culture. In this way, an artificial early stage—a blastocyst-like blastocystoid (or blastoid) can be obtained, and its further development investigated.

Cells involved can even be genetically modified and influenced in various ways. The blastoid continues to develop *in vitro*, giving rise to a "synthetic" embryo with the rudiments of a neural tube, intestine and limbs, and a beating heart (Amadei et al., 2022; Tarazi et al., 2022). The benefits that accrue from this procedure are straightforward. In addition to the fact that larger amounts of samples can be prepared, there is also the possibility of manipulating the composition of the media, the timing of the different agents, the choice of the source of the three cell types, and the introduction of mutations and gene markers where appropriate. Whereas experiments on human embryos are drastically restricted by legislation, there are no objections to experiments on tissue cultures or to the construction of organoids; and the blastoid is a similar model, it was not formed by the fusion of sex cells and cannot give rise to a human being. It can therefore be used to study— in a controlled way—the various subtle *partial* regulations of developmental processes.

Close Cohabitation This section could also belong to the following category, yet we exclude it from "ecology in general" because we want to draw attention to the close, directly intimate cohabitations of distinct and unrelated species. The reader will certainly think of the host-parasite relationship, where the parasite often attacks some specific host species only; moreover often cycles between different host species: for example, a mammalian—insect vector in the case of malaria, sleeping sickness or tapeworms, etc.

Metabolic consortia of bacteria and archaea are also ubiquitous. An example is *Prometheoarchaeum syntrophicum*, an archaeon which we mention in a different context below: cultivation of this archaeon only succeeds in co-culture with the bacterium *Methanogenium* (Imachi et al., 2020). Similar cases, however, may involve not only pairs, but perhaps dozens of microorganisms from different communities with variable species representation, and thus belong rather to the following section.

The relationship between plant roots and the fungus mycorrhiza is a very close symbiosis (see e.g., Sheldrake, 2020; Markoš & Švorcová, 2019). In particular, the fine hyphae of arbuscular mycorrhiza penetrate deep into plant cells, but both forms of Closure 1—plant and fungal—retain their autonomy, although both undergo extensive differentiation processes due to symbiosis (Rasmussen et al., 2016). Moreover, fungal filaments grow through the environment over long distances and can interconnect not only metabolically but also informationally with perhaps all plants living in a given community (e.g., forest or meadow; Klein et al., 2016). Thus, the whole of the community can coordinate its responses to different environmental challenges (pathogen, herbivore, chemism, etc.).

Ecosystems Examples of the coexistence of organisms on a larger scale include different mats, stromatolites, holobiotic relationships, a forest, a meadow, or marine and ocean habitat communities. They are characterized by elaborate relationships and structures, and may bring together tens or even hundreds of different species. Individual species of cells or multicellular bodies or consortia enter such relationships, but may also exit them, or at least send their propagules beyond the community. The list of species or populations that make up an ecosystem is neither finite nor permanent, but changes over time, or in according to the season or the presence or absence of some component (e.g., calamity from overpopulation, epidemics, size of the area, climatic factors such as season or flooding, etc.). The inhabitants of an ecosystem significantly and extensively coordinate their relationships with each other. So again: they need to know each other and interact in a meaningful way.

A special kind of ecological symbiosis is the *holobiont*, a macroscopic organism such as the human body, which is "inhabited" by hundreds of other symbionts; their species composition can change, hence the "fluctuating chimera" (Chiu & Gilbert, 2015). A full-fledged organism is nearly always a holobiont; axenic (non-microbial) organisms generally cannot survive in isolation (see Sect. 8.6) and such "germ-free" organisms must be laboriously maintained in the laboratory. Human newborns with lethal immune deficiency must be removed by caesarean section and live—often for years—in a sterile tent devoid of microbes until a suitable bone marrow donor is found for them.

Mammals begin to recruit their symbionts already during the passage through the birth canal and during suckling, by licking objects, inhalation, etc. Similarly, plant roots are constantly organizing their *rhizosphere* from soil bacteria, fungi, protists—as well as from the roots of other individuals. On this "scaffolding" of extracellular matrix, other symbionts—presumably in a controlled manner—mount. The fact that full-fledged ontogeny is almost always a matter of symbioses with other species (and plastically adjusts according to the environment offered by the symbionts) has even led to the establishment of ecological evolutionary developmental biology (eco-evo-devo; Gilbert et al., 2015).

Cell Fusion and Intracellular Symbiosis In countless intracellular symbioses, a kind of "nesting" occurs: though the endosymbiont inhabits the host's cytoplasm, and thus has overcome its Closure 1, but it retains its own Closure 1. Often, the symbiont enters the next generation of host cells directly, established in the cytoplasm (e.g., *Chlamydia*, *Wolbachia*, *Rickettsia*), but the possibility remains for it to leave the host cell and infect a new host.

Some protist cells are tightly packed with endosymbionts: think of ruminant rumen ciliates, or polychaete protists (Polymastigina) from the gut of termites with methanogenic archaeal symbionts in their cytoplasm. Nitrogen-fixing bacteria of the genus *Rhizobium* or *Frankia* multiply in the cells of root tubers. In specialized organs of insects (mycetomes, bacteriomes) we find symbiotic fungi or bacteria that help their host digest often very unusual types of food (and thus underlie the unusual success of insects at occupying a wide variety of niches; review e.g., Markoš & Švorcová, 2019; Bourtzis & Miller, 2003). In the cytoplasm of some ciliates, but also in the cells of cnidarians (hydra, corals) or gastropods, we find symbiotic photosynthesizing cyanobacteria or algae that can often turn their host into—technically speaking—a plant, without the need to ingest any food.

When the sex cells fuse, a new Closure 1—the zygote—is formed. However, the word "formed" is used with embarrassment: it did not come into existence, as with every Closure 1 it has always been there. Nor is it a symbiosis: the fusion has brought about a qualitative reorganization of the egg cell interior. But the fusion of Closures 1 is not limited to sexual processes. At the site of a graft, the plant cells of two different organisms fuse, with a subsequent fusion of the nuclei; this may serve as an example. Such cells can give rise to a viable shoot, and sometimes even a new species of organism (Fuentes et al., 2014). Similarly, mycorrhizal fungi (e.g., *Glomus*) have syncytial hyphae with nuclei of different origins; the "propagulae" in these fungi are formed by fusion of hundreds of hyphae into a syncytium containing thousands of nuclei; this then gives rise to hyphae with new combinations of nuclei (Genre, 2012; Moore, 1998).

Symbiogenesis Above we discussed endosymbiosis and emphasized that an endosymbiont is a full-fledged cell that can exit the host cytoplasm. However, in very rare—but from an evolutionary point of view very important— cases, the endosymbiont gives up its closure and turns into a *semiautonomous organelle*, a chimera of Closures 1, at which point the separation of the two partners is no longer possible. Several steps had to be fulfilled for this to happen (see e.g., Lane, 2015).

According to the present state of knowledge, the eukaryotic cell itself is the product of a symbiogenetic event, when the cell of an archaeon fused with a bacterium. The palimpsest of this event in today's eukaryotic cells is the genetic apparatus of the archaeal type, while the metabolism and membrane structures are of the bacterial type. It was even possible, after many years of effort, to find and grow in the laboratory a species of archaeon from a candidate group (*Prometheoarchaeum syntrophicum*). Characteristically, the cells have long "tentacles", so that the discoverers of the species immediately came up with a new theory of the origin of the eukaryotic cell, which they called "E^3": *embrace—engulf—endogenize* (i.e., integrate into your cytoplasm; Imachi et al., 2020).

Proto-eukaryotic cells later apparently received another bacterial endosymbiont (α-proteobacterium), which gradually evolved into mitochondrion, an organelle that provides respiration as well as synthesis of ATP and many other substances. After eukaryotes (already equipped with mitochondria) diversified into many developmental lineages, one of these underwent a similar event: a photosynthesizing bacterium (cyanobacterium) was transformed into an organelle, and these cells gave rise to the lineage of plants (Archaeplastida). This event led to a tremendous increase in the efficiency of photosynthesis in the biosphere and this group feeds most of today's biosphere; it also produces oxygen as an oxidant for the respiration of other groups of aerobic life.

However, symbiogenetic events, rare as they are in evolution, are not yet over: in other branches of eukaryotes, many other groups can also photosynthesize, however (with one known exception) they did not start with cyanobacteria but turned already established eukaryotic algae from the Archeplastida group into organelles. This process even recurred two or three times in some groups!

10.6 Evodevo

"Contrary to the expectations of *any* biologist, most of the genes first identified as governing major aspects of fruit fly body organization were found to have exact counterparts that did the same thing in most animals, including ourselves", states evolutionary developmental biologist Sean Carroll (2006) in the introduction to his book. It turned out that the "toolkit", i.e., the proteins controlling the early ontogeny of an individual (and thus also the set of genes that some biologists believe "control" it all), is highly homologous in large groups of life, such as all vertebrates, all animals, all multicellular eukaryotes, etc.

A definitive (though sometimes circumvent) rule of development in a *phylum* is the so-called *phylotype*, an early embryonic stage of broadly similar appearance for all lineages of that phylum (in the case of chordates, from the larvae of tunicates through the tiny fish and hummingbirds to brontosaurs and whales; similarly, the phylotype in vascular plants, insects, or fungal fruiting bodies). The phylotype of chordates is the so-called *pharyngula*, an articulated embryo with a dorsal string (notochord), an anteriorly open nerve tube without a brain but with the bases of the head sensory organs (placodes), pharyngeal (hence pharyngula) slits, and limb buds—the fundamentals of the appendages. The digestive system tends to open into the extraembryonic parts of the body (yolk sac, placenta, etc.) used for nutrition and excretion. Every schoolchild is familiar with the pharyngula, but it is rarely pointed out that the paths leading from the zygote to this common stage are many: the giant ostrich egg develops differently from the tiny human egg—yet in both cases the same "toolkit" is used, only the initial structures are built differently with the help of the tools. This is where the concept of "epigenetics" began—the building of structures that depend not only on the quality of the genetic information, but also on the quality of the cytoplasm of the egg (and its surroundings). The epigenetic stages of the embryo in different groups were well described in the nineteenth century, hence the surprise when it turned out that members of a species or lineage use the same genetic tools to build many quite different structures, only to end up—with the phylotype.

From the stage of the phylotype onwards, the ontogeny of the embryo proceeds according to the principles established in the given "classes" (differently in mammals, birds or frogs) and with time it increasingly takes the Appearance of a member of a genus or species. Thus, developmental evolutionary biology (evodevo) can also be understood as an exploration of the dialectical relationship between tradition (conservative memory, long-established sets of tools, repetitive structures of body plans) and the present, during which tradition is *somehow* interpreted to suit the species-specific conditions into which the individual was born. And since, as we have pointed out several times, environmental conditions are mainly set by the other inhabitants of the biosphere, we also come to the idea of "eco-evo-devo". The "here and now" takes into account the open future, but also carries with it (or uses) the burden and experience of tradition.

Tradition (the "collective" memory) also carries with it a communicative component, and here another kind of epigenetics comes into play—those transmitted from parents to zygote, those transmitted in some lineages from parents to offspring through learning, and those that the individuals accumulate themselves, some of which they can then pass on to their offspring. We shall focus on learning and teaching while concentrating on the communication between the closures.

10.7 Inheritance of Acquired Characteristics

In this subsection, we stick mainly to the monograph *Extended Heredity. A New Understanding of Inheritance and Evolution* (Bonduriansky & Day, 2018). We have already discussed epigenetic modifications at different levels above. We have also seen that they are behind the determination and differentiation processes within cell lines. The norms of reaction and other forms of plasticity are also epigenetically based. Perhaps it will not surprise the reader that some of these modifications can be transmitted through sex cells to offspring and their progeny without being fixed by mutation (for some cases, see Markoš & Švorcová, 2019; Švorcová, 2023). Bonduriansky & Day (2018, p. 7) note that an individual may even exhibit traits that do not result from the genotype it inherited but from the genotype of its parents—for example, when a daughter (mouse) carries a trait encoded by the father's Y chromosome (Nelson et al., 2010). There are thus multiple channels for the transmission of inherited traits, hence the term "*extended heredity*" —for example, Danchin (2013), Danchin et al. (2011) and Danchin and Wagner (2010). There is also increasing evidence for the influence of hormone analogues and other regulatory substances, even via sperm or seminal fluid, as is the effect of vincosolin A (an anti-androgen), in Schuster et al. (2016).

And so "Biologists are now faced with the monumental challenge of making sense of a rapidly growing menagerie of discoveries that violate deeply ingrained ideas. One can get a sense of the growing dissonance between theory and evidence by perusing a recent review of such studies and then readin: 7). Is there a shortcut to the somatization of culture looming here?

References

Ahnert, S. E., et al. (2015). Principles of assembly reveal a periodic table of protein complexes. *Science, 350*, aaa2245.

Amadei, G., et al. (2022). Embryo model completes gastrulation to neurulation and organogenesis. *Nature, 610*, 143–153.

Augustine, A. (2021). *[426] The city of God, 1*. T & T Clark.

Barfield, O. (2013). *[1977] The rediscovery of meaning and other essays*. Barfield Press.

Biller, S. J., et al. (2014). Bacterial vesicles in marine ecosystems. *Science, 343*, 183–186.

Boddy, A. M., et al. (2015). Fetal microchimerism and maternal health: a review and evolutionary analysis of cooperation and conflict beyond the womb. *Bioessays, 37*, 1106–1118.

Bonduriansky, R., & Day, T. (2018). *Extended heredity. A new understanding of inheritance and evolution*. Princeton University Press.

Bourtzis, K., & Miller, T. A. (Eds.). (2003). *Insect Symbiosis*. CRC Press.

Cai, Q., et al. (2018). Plants send small RNAs in extracellular vesicles to fungal pathogen to silence virulence genes. *Science, 360*, 1126–1129.

Carroll, S. B. (2006). *Endless forms most beautiful: The new science of evo devo*. Norton & Co..

Chiu, L., & Gilbert, S. F. (2015). The birth of the holobiont: multi-species birthing through mutual scaffolding and niche construction. *Biosemiotics, 8*, 191–210.

Chivian, D., et al. (2008). Environmental genomics reveals a single-species ecosystem deep within Earth. *Science, 322*, 275–278.

Chuong, E. B. (2013). Retroviruses facilitate the rapid evolution of the mammalian placenta. *BioEssays, 35*, 853–861.

Danchin, É. (2013). Avatars of information: towards an inclusive evolutionary synthesis. *Trends in Ecology and Evolution, 28*, 351–358.

Danchin, É., & Wagner, R. H. (2010). Inclusive heritability: combining genetic and nongenetic information to study animal behavior and culture. *Oikos, 119*, 210–218.

Danchin, É., et al. (2011). Beyond DNA: integrating inclusive inheritance into an extended theory of evolution. *Nature Revviews Genetics, 12*, 475–486.

Dasmahapatra, K. K., et al. (2022). Butterfly genome reveals promiscuous exchange of mimicry adaptations among species. *Nature, 487*, 94–98.

Denton, M. J., et al. (2002). Protein folds as Platonic forms: New support for the pre–Darwinian conception of evolution by natural law. *Journal of Theoretical Biology, 219*, 325–342.

Diamond, S. E., & Martin, R. A. (2021). Buying Time: Plasticity and Population Persistence. In D. W. Pfennig (Ed.), *Phenotypic plasticity and evolution. Causes, consequences, controversies* (pp. 185–209). CRC Press.

Dorit, R. L., et al. (1990). How big is the universe of exons? *Science, 250*, 1377–1382.

Edelman, N. B., et al. (2019). Genomic architecture and introgression shape a butterfly radiation. *Science, 366*, 594–599.

Fuentes, I., et al. (2014). Horizontal genome transfer as an asexual path to the formation of new species. *Nature, 511*, 232–235.

Gabius, H. J. (2018). The sugar code: Why glycans are so important. *BioSystems, 164*, 102–111.

Genre, A. (2012). Signalling and the re-structuring of plant cell architecture in AM symbiosis. In S. Perotta & F. Baluška (Eds.), *Signaling and Communication in Plant Symbiosis* (pp. 51–71). Springer.

Gilbert, S. F., et al. (2015). Eco-Evo-Devo: developmental symbiosis and developmental plasticity as evolutionary agents. *Nature Reviews Genetics, 16*, 611–622.

Gjorevski, N., et al. (2022). Tissue geometry drives deterministic organoid patterning. *Science, 375*, aaw9021.

Halbwachs, M. (1980). *[1950] The collective memory*. Harper & Row.

Hall, B. K. (1999). *The neural crest in development and evolution*. Springer.

Harmston, N. (2020). Regulation in common: sponge to zebrafish. *Science, 370*, 657–658.

Imachi, H., et al. (2020). Isolation of an archaeon at the prokaryote–eukaryote interface. *Nature, 577*, 519–525.

Kalluraya, C. A., et al. (2023). Bacterial origin of a key innovation in the evolution of the vertebrate eye. *PNAS, 120*, e2214815120.

Kauffman, S. A. (2000). *Investigations*. Oxford University Press.

Klein, T., et al. (2016). Belowground carbon trade among tall trees in a temperate forest. *Science, 352*, 342–344.

Kyprianou, C., et al. (2020). Basement membrane remodelling regulates mouse embryogenesis. *Nature, 582*, 253–258.

Lane, N. (2015). *The Vital Question: Why is life the way it is?* Profile Books.

Li, Y., et al. (2022). HGT is widespread in insects and contributes to male courtship in lepidopterans. *Cell, 185*, 1–13.
List, J. M., et al. (2013). Networks of lexical borrowing and lateral gene transfer in language and genome evolution. *Bioessays, 36*, 141–150.
Markoš, A. (2002). *Readers of the book of life. Contextualizing developmental evolutionar biology*. Oxford University Press.
Markoš, A., & Švorcová, J. (2019). *Epigenetic processes and the evolution of life*. CRC Press.
Melentijevic, I., et al. (2017). *C. elegans* neurons jettison protein aggregates and mitochondria under neurotoxic stress. *Nature, 542*, 367–371.
Moore, D. (1998). *Fungal morphogenesis*. Cambridge University Press.
Nelson, V. R., et al. (2010). Transgenerational genetic effects of the paternal Y chromosome on daughters' phenotypes. *Epigenomics, 2*, 513–521.
Nietzsche, F. (1997). [1873–1879] On the uses and disadvantage of history for life. In *Untimely meditations* (pp. 57–154). Cambridge University Press.
Papale, F., et al. (2020). Networks consolidate the core concepts of evolution by natural selection. *Trends in Microbiology, 28*, 254–265.
Pennisi, E. (2016). Shaking the tree of life. *Science, 354*, 817–821.
Rasmussen, S. R., et al. (2016). Intraradical colonization by arbuscular mycorrhizal fungi triggers induction of a lipochitooligosaccharide receptor. *Scientific Reports, 6*, 29733.
Raup, D. M. (1991). *Extinction: bad genes or bad luck?* Norton & Co.
Revah, O., et al. (2022). Maturation and circuit integration of transplanted human cortical organoids. *Nature, 610*, 319–326.
Schuster, A., et al. (2016). Ancestral vinclozolin exposure alters the epigenetic transgenerational inheritance of sperm small noncoding RNAs. *Environmental. Epigenetics, 2*(1), dvw001.
Sheldrake, M. (2020). *Entangled life*. Penguin Random House.
Snetkova, V., et al. (2022). Perfect and imperfect views of ultraconserved sequences. *Nature Reviews Genetics, 23*, 182–193.
Suzuki, T. (2021). The expanding world of tRNA. *Nature Reviews Molecular Cell Biology, 22*, 375–392.
Švorcová, J. (2023). Transgenerational Epigenetic Inheritance of Traumatic Experience in Mammals. *Genes, 14*, 120.
Švorcová, J., & Markoš, A. (2023). Closures as a Precondition of Life, Agency, and Semiosis. *Biosemiotics, 16*, 45–59.
Tarazi, S., et al. (2022). Post–gastrulation synthetic embryos generated ex utero from mouse naive ESC. *Cell, 185*, 3290–3306.
Toyofuku, M., et al. (2023). Composition and functions of bacterial membrane vesicles. *Nature Reviews Microbiology, 21*, 415–430.
Wilmut, I., et al. (1997). Viable offspring derived from fetal and adult mammalian cells. *Nature, 385*, 810–813.
Wong, E. S., et al. (2020). Deep conservation of the enhancer regulatory code in animals. *Science, 370*, eaax8137.

Chapter 11
Analogies

Abstract Is an analogy allowed between the narrative schemes of historians and those of evolutionary biologists? Throughout this book, we answer "yes", arguing in favor of the statement. Here a short summary. (1) All living beings are characterized by their innerness that allows interpreting their situatedness in the world, here and now. We suggest differentiate inner processes into two Closures, characterized as cell interior (Closure 1) and cell communities (Closure 2). (2) For evolution of any community it is typical to start with a short plastic phase when many alternative paths to the future are possible. It will follow with the elasticity phase when the community becomes more conservative and tends buffering all deviations back to "norm". (3) We introduced the concept of Appearance, as a sum total of all possible phenotypes of the living being. Its evolution proceeds through evolution of phenotypes that can be found in different phases on the plasticity–elasticity scale. (4) Finally, we express some questions, even doubts, that will need further thinking and research.

Keywords Innerness · Plastic and elastic phases · Analogies · Culture

> Perhaps it goes without saying that it is important to avoid any hint of vitalism. There is no reason to assume that organisms are endowed with an inherent ability to respond in a beneficial way to perturbations and stresses. It is true that most mutations that affect fitness are deleterious, as are many or most environmental changes. (The vast majority of species are no longer here.) But the plastic compensatory mechanisms of individual organisms that mitigate such deleterious effects are not inherent in living things; they are the result of evolution. Thus, if the capacity for phenotypic plasticity provides the basis for a favorable, genetically established trait, this plasticity itself requires an explanation—and the explanation is often, or usually, our old friends—genetic variation and natural selection. Many non-biologists, however, may prefer theories such as Lamarckism in place of natural selection.
>
> (Douglas J. Futuyma, 2021, p. 360)

A. Markoš, J. Horský, *The Seeds of Forms and the Forms of Seeds*, Biosemiotics 12, https://doi.org/10.1007/978-3-032-16131-4_11

As our text draws to a close, the question returns over and over again: can we draw an analogy between the narrative schemes of historians and those of evolutionary biologists? In many places we argue that we can, and yet a careful reader may have detected some embarrassment at this answer.

Our second question is even more beseeching: We may well have defined life as a semiotic category, and yet it seems that we feel that human life, culture or history are somehow *different from* what one would expect from other kinds of life, from their relationship to the world and the story of their evolution. Can it be that a bacterial mat, a philosopher, an embryo, and a mob of football fans all belong in the same category? Yes, from the perspective of this book they do, and yet... Let us try once again, in conclusion, to identify some neuralgic points of embarrassment.

11.1 Innerness

We introduced Darwin as a proponent of Newton's doctrine (Chap. 4), noting that to incorporate biology into Newtonian physics is to deny innerness to the living and describe organisms as any other body moved by external forces. Here's yet another quote: "If Darwin's theory was to pass muster with the likes of Herschel and Lyell, who demanded theories that looked as much as possible like Newton's, in which matter was to be more acted upon than acting, Darwin would have to find an external force, like Newtonian gravity, rather than internal drive, that impinged on the development and reproductive cycle with sufficient force to drive and shape evolutionary diversity. That force must, moreover, be a *vera causa*, a real cause that can be seen at work this very day" (Depew & Weber, 1996, pp. 70–71). This tradition is still held sacrosanct by most evolutionary biologists today, see below.

There is yet another potential transgression that biologists are said they must avoid: "The fact that it does no good to speculate about what might have happened in bygone ages is underlined by the fact that such speculation always takes the form of a narrative. If the aim of science is to see how much one can explain by rules governing observable phenomena, however, a scientist must avoid giving up the search prematurely simply by postulating ad hoc, and question begging, stories that could have accounted for the occurrence of something long ago. Science is not narrative. As in physics, the rules or laws governing changes in the earth must be presumed to be the same throughout time. What goes on in the world, therefore, is an endless series of law-governed instances rather than a concatenation of episodes" (Depew & Weber, 1996, p. 97; see also Sects. 4.5 and 8.3). As we have seen in Chap. 4, even Darwin himself sinned against this behest; in the end, evolutionary events are necessarily interpreted as a story.

In this book, we note that living—hence semiotic—beings have an innerness, and that this allows them to evaluate everything that fits within their *space of experience* (which is identical to their umwelt (Markoš & Švorcová, 2019; Švorcová & Markoš, 2023) and to schedule their progress into the future, to choose an azimuth to a certain region on the horizon of expectations of things to come. Moreover, contingencies *outside* the space of experience allow for abduction, a creative approach that can lead to an expansion of this space and a preservation of the viability of the individual and/or community. Or to a collapse of such an enterprise: with the scheduling of the future, we always go out on a limb.

Yet even today, we mostly encounter an emphasis on scientific knowledge without recognition of the innerness of the objects studied. In contrast with this more common conception of evolution, we choose for illustration a passage where evolution is characterized by J. Flegr. He states: "Evolution is generally understood to mean gradual development of any system with a 'memory', i.e. any system that responds to external stimuli due to the stimuli that it encountered in the past. This means that it is equally possible to speak of the evolution of languages, automobile chasses or ladies' hairstyles as about the evolution of conifers (Flegr, 2021, p. 2).

While such a broad conception of "evolution" is surely possible (and colloquial English allows it), it prevents any further investigation: it simply says that things change over time and past states affect the present; this is true perhaps of any and every thing in the universe. We can, for example, describe the meandering of a river or the deflection of the earth's axis in this way; indeed, all historical events in the world will again fall into the same category. We argue, therefore, that memory is not enough: a living being is necessary wherein memory is, first, registered, and, second, joined by the experience necessary for interpreting the world. Experience does not necessarily have to take the form of a linear molecule of nucleic acids: all past structures or artifacts that alter the environment of the "system", which is the entire setup of experience, also play a role in how the Appearance *interprets* its momentary existence. When we speak of evolution in this book, we understand it in this way—and only in the context of the history of life. This brings us to the platform on which this book is built, and which has perhaps freed us from the constant defenses against the charge of anthropomorphism, determinism, and similar heresies against good scholarly manners.

We have built our text on the assumption that evolution is a subtle game between Closure 1, represented by cells, and Closure 2, represented by communities of cells in various forms: from the symbiotic relationship between two cells to evolutionary developmental biology and human civilizations. This game creates pictures of the world that are valid at a certain point in time. Such communities, like cells themselves with their spaces of experience and horizons of expectation, can also be highly variable and range from ad hoc settings for varying lengths of time, to communities with established rules (e.g. multicellular bodies, mycorrhizae, or social insects). Closure 2 is more about understanding the cohabitants, whether they belong to our kin or to the category of "others". Such an understanding undoubtedly has its evolution based on cells (Closure 1) but is constituted on different principles—precisely on that understanding. The reader may wonder where we left the "environment", which alone is, according to many theories, crucial for natural selection, to determine what further paths evolution will take. We judge that for the most part it is represented by other Closures (1 or 2), this being the active component that makes and interprets decisions; the physical environment (meteorite fall, temperature, position of continents, seasonal events, etc.) represents the scenery, the boundary (playground, constraints) that must be respected or, in some cases, be abolished or modified (we do not claim, of course, that this is always by active intention and effort: natural selection has a say here).

Yet it is often assumed that human communities are somehow completely different from the Closures of other lineages: after all, only humans think and talk! This is where sophistry enters; based on the assumption of different types of semiotics, the biosphere is divided into one "higher" and many "lower" species. We have tried to cleave to the assertion that the semiotic endowment of organisms is of one type only, with different accents (and outcomes). And we have also tried to show that the "syntax" of semiotic relationships in all inhabitants of the biosphere is built on the same principles, just as their very existence is built on very similar genetic, metabolic, and signaling systems.

What is even worse, many researchers—whether biologists or historians—deny the innerness of other lineages, and (except perhaps in some species of "higher" animals) would prefer to relegate non-human beings to the role of Cartesian machines. From this view, everything said above will be met with a dismissive response. It is beyond doubt that experimental biology as well as experimental medicine derived from it, must often work with such a Cartesian assumption, otherwise science wouldn't be possible. But the experimenters know, or should know, the axioms with which they work (see the quote from Futuyma in the epigraph above).

11.2 Plasticity, Elasticity, Hypolepsis

Before we continue, a small digression to the theory of frozen plasticity by J. Flegr (2010, 2021, also Toman & Flegr, 2018), and with an interpretation of the basic theses as described by Markoš & Švorcová (2019). While Toman and Flegr concentrated on sexually reproducing organisms, we argue that the curves in Fig. 11.1 may reflect a general trend of evolution of lineages, in both life's closures.

The main curve of the diagram in Fig. 11.1a explains the basic principle of the evolution of a lineage, such as a species. More precisely, it shows the evolution of one of the many phenotypes that we can define on members of a lineage (species, culture, language), so the curve represents only a slice of the multidimensional space of phenotypes (in the case of the grouses, as an example, feather texture or color, stomach structure, sexual dimorphism, or wedding ritual). Thus, the horizontal axis of the graph shows time (more precisely, the number of generations), while the vertical axis shows the evolution of one selected phenotypic parameter.

Moreover, the curve is distorted for the sake of clarity: its ascending, *plastic* phase actually occupies only a small fraction of the "lifetime" of the lineage; the *elastic* phase, during which the characteristic of the parameter under study is dominant. In the plastic phase, the trait under study can diverge often and drastically, giving rise to new forms or even species; this is illustrated by the arrows diverging from the "main" line (i.e., picking a different trajectory). With time, the links between the various traits in the population become more and more established and

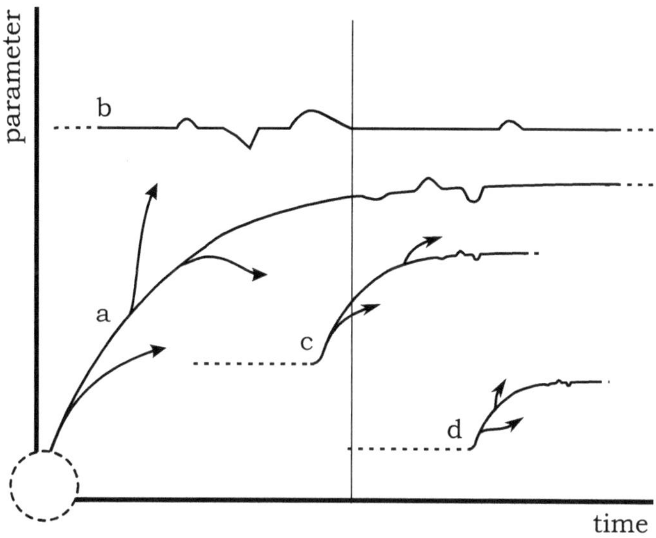

Fig. 11.1 *Plastic and elastic phases in the evolution of phenotypes.* Note that that the axes have no beginning: there has always been something before. Several graphs ("transparencies") are laid over each other. (**a**). A new feature whose evolution "fits" the graph with its plastic and elastic phases (e.g., the evolution of the forelimbs of cetaceans). (**b**). The wavy, nearly horizontal line is some ancient, elastic feature (e.g., early vertebrate embryogenesis). (**c–d**). Imagine an overlap of dozens or hundreds of similar curves stacked, each with a different onset and progression of its plastic or elastic phases. Moreover, there is the possibility of the discovery of novelty, or the plasticization of a feature, at different times. This is illustrated in two instances of such "explosions" flashing from the back rows (examples: a new symbiosis within the holobiont; adaptive radiation or the sudden discovery of the possibilities of a norm of response; mind and speech; hybridogenesis; immunization; a deflection in umwelt space to different coordinates, and a new interpretation of what is there, etc.). The Appearance of the individual or species is then given, at a given moment, by a perpendicular cut through the tangle of all those curves structuring its overall "non-contemporaneous contemporality", multidimensional, each of the dimensions having a different dynamic at any given moment

it becomes increasingly difficult to change their interrelationships; the population enters the elastic phase of its history. If the external pressure becomes significant in this phase, the parameter may change, only to return to the original values or close to them after the pressure has ceased.[1] A population (of a species) persists for an average of five million years before it becomes "ossified" to such an extent, that it fails to respond to any of the environmental challenges and either becomes extinct, or a catastrophe disrupts established "habits" to such an extent that the population (or rather a fraction of it) returns to the plastic phase, and we observe the emergence of a new species. To paraphrase some of the conclusions from Markoš & Švorcová

[1] The established parameters of human culture are dislocated quite considerably, for example, by war or epidemic, yet after the disaster is over, society recovers quite quickly—despite the belief that "nothing will be as it was".

(2019, pp. 185–188), in the plastic stages of the development, the Appearances are 'tests' of different forms of Closure 2, which discover "what is allowed" or "acceptable" under new conditions; moreover, because some rules are still flexible, a plethora of possible trajectories will emerge. In later phases, the rules begin to "freeze", and more stable heuristics emerge to bring the community together, stabilize it, and 'build in' time saving routines. The new variant stabilizes as a new point against a much broader backdrop of umwelt possibilities. These represent the memory and experience of individuals in the community, called "norms". An imposing analogy between otherwise diverse communities of thinkers (Flegr, 2010; Lotman, 2009, 2010; Rappaport, 2010) thus points to the obvious fact that in their beginning interpretations of the world must be negotiated within frames of such a community, whereas later "habits" or "norms" would prevail. Abrupt turns (catastrophes, "explosions") allow a community to (re)construct anew its collective past and (if successful) take on a new plasticity. In other words, communities interpret their past (led by mythology, hypolepsis, current needs etc.) to *invent their present anew*.

The elastic phase, on the other hand, represents a kind of a *steady state*, which is constantly challenged by environments that cause deviations large or small. However, as we noted above, different elements of the relations in the two Closures occur at different stages of evolution. Ideally, perhaps, only a small part of phenotypes is in the plastic phase, with other elastic traits buffering the survival. Some traits may even be in the decay stage of the elastic phase, a crisis stage that may threaten the very existence of the lineage, or conversely lead to the elaboration of new plastic strategies. Since the beginning of evolution, the following two strategies have led to a tipping of the system into a different regime (in the way of a kaleidoscope), or to novelties.

1. Random apomorphies—these are not really strategies at all, simply the existing configuration of plastic and elastic processes that allow the incorporation (i.e., interpretation) of novelty (or even a new regime) into the form of a "system" —it doesn't matter whether the novelty is a catastrophe brought by the environment or by a mutation. Evolution is triggered by the embodiment of new variants as mutation, horizontal transfer, intrusion or hybridogenesis, i.e., qualitative changes of avatars or parts of avatars.
2. Delegation of the problem to another Closure or to other elements of the Closure, with the understanding that it may tip over into a new mode, a new space of operation. Humans, for example, are elastic in body structure (even if only to a very limited extent) but are highly plastic in their ability to learn and cooperate within various newly established cultural and perhaps epigenetic Closures, tolerance to different ways of living (as, e.g., types of diet, the presence of symbionts, established ways effective in the society, etc.). Such a property is also possessed, for example, by experienced cultures, which respond plastically (often surprisingly) to circumstances; for the resilience of Latin culture over the centuries, see, for example, Deely (2010) or Gurevich (1990).

One more peculiarity of the theory of Toman and Flegr (2018): they assume evolution based on sorting in terms of stability, as the properties of organisms stabilize over time, deviations from stability become smaller and more trivial. This is a remarkable insight. In Sect. 8.3, we report that the spontaneous decay of the universe into a state of minimum internal energy and maximum complexity is considered by many to be a form of "evolution", albeit a rather bizarre one. Here we are dealing with evolution to maximum complexity (whatever the criterion of complexity might be), with an increasingly sophisticated *structuration* of that complexity; even so, biological evolution seems to be moving (accordint to authors) towards a steady state and then ceasing. Fortunately, the environment changes much faster than the progression to stability, and so the curve is constantly being deflected from its "normal" course: plastically, elastically, by extinction, or by the initiation of a new plastic–elastic cycle.

Toman and Flegr (2018), in our opinion, preferrably see the basic level of description in the genotypic makeup of the population: the frequency of different alleles (as in the bacterial population in Sect. 8.6), epistatic interactions and regulation, etc. The parameter measured on the vertical axis (Fig. 11.1) can, for example, be the number of alleles of a certain gene in the population, or their (somehow defined) interactions, or also the phenotype, which as a function of genotype *is* in this case identical to the Appearance. Under this assumption, then, the model can indeed be represented by a single curve, but this—if we take into account the plethora of other phenotypes—can function as a *pars pro toto* for describing the general history of life on the planet. Indeed, all aspects of life: the curve also fits the evolution of human cultures (Lotman, 2010) or mythologies (Rappaport, 2010); for extensions to all aspects of the Appearance see e.g. Danchin (2013). Danchin likens genetic notation to an "avatar" that represents the most conservative, evolutionarily stabilized component. However, an avatar by definition requires embodiment, and here begins the play of corporeality, innerness, habituation, etc., that allow the realization of the avatar's "message". The author then defines a range of interactions, some of which are inherited (with varying degrees of accuracy), but others are only passed on to the next generation through learning. We also work with the assumption that cultural phenomena typical of humans represent only special cases of the general rules of the evolution of life, cases only made possible by thought and speech (also Markoš & Švorcová, 2019, p. 185).

To capture all forms that enable the transmission of the memory and experience of Appearances in the biosphere, Danchin (2013) proposes a scale from "hard" (easy to study by scientific methods) genetic inheritance to increasingly "softer" traits achieved through epigenetic, cultural, parental and temporal context–derived forms, etc. Somewhat aside is the "superstructure" of "exo–somatic" non–genetic tools of transmission that takes the form of writings, engineering and architectural works, perhaps including landscape transformations, and are not necessarily carried out by humans alone. In the opposite direction is genetic memory, most conservative in the harder forms of transmission. This is, again, hypolepsis coupled to novelty and new interpretations.

If we consider the individual curves in Fig. 11.1, we can take each as one facet of a multidimensional hyperspace of phenotypes. This offers another possibility: the evolution of different phenotypes of a single form may not be synchronized, just as, by analogy, the ontogeny of the Appearance of an individual is not synchronized, as is case of an overlap of generations. Some phenotypic aspects may be frozen for good, even as others enter a plastic phase or, conversely, are frozen only for a time.

Let us take the giraffe as an example again: its early embryogenesis is hardly distinguishable from similar processes in vertebrates, mammals, or artiodactyls—hence the body segments preserve the branching of, for example, the brain nerves, which results in an "illogical" innervation of the thyroid gland. The structure of the ruminant digestive system does not differ from other ruminants (except for the length of the esophagus), and the hoof structure is likewise not innovative. On the other hand, the general body habitus as well as some physiological adaptations (such as the vascular system) are genus-specific and have undergone considerable change.

Cichlids are also a good example of asynchronous lineage evolution. Both these examples highlight the incorrectness of the curve in Fig. 11.1a, which after all, does not start at point "zero". We emphasize the trivial fact of the non–existence of a "point zero" because even inanimate dissipative structures can have a similar evolution described by our curve. These, however, do have a "true" origin, emerging repeatedly de novo from differently organized matter: a hurricane or a flame, for example. Thus, in Fig. 1.1 we drop the origin, and superimpose a bunch of curves, each for a different phenotype and with different dynamics of onset and evolution. The reader will find a more detailed description of the figure—perhaps the most important summary of our text—in the legend.

11.3 Doubts

But let's go back to the start where we said that the main axiom (belief) of most biologists is the assumption that living things are *objects* that are no different from inanimate objects in physics—hence a belief in a basic level of description that determines everything. A good illustration of this is the conclusion of Chap. 9, where, among other things, we draw from a recent collection of papers on plasticity and the norm of reaction (Pfennig, 2021). The topics presented offer a range of interpretations, yet the individual authors do not fail to recall their belief which can be summarized by the statement "But in reality...". The evolutionary biologist Futuyma expressed this most elegantly in his paper (see the epigraph to this chapter). He may well be right—most evolutionary research programs work with such axioms and produce fantastic data. But the data may be subject to other interpretations within different paradigms—that's what we were trying to point out. Moreover, every scientific investigation also presents projects of desire that necessarily alter and distort the object under investigation. We are not immune: we have searched for sources, we have dreamt up stories, and we have also allowed ourselves a whit of rhetoric.

And questions and doubts arise—we have dealt with them; here is just a summary:

1. The only way we can proceed is by the *analogies* we can draw between biological and historical thinking, or between the evolution of life and the development of (human) culture. The only exception is perhaps the possibility of asking whether culture (or its more autonomous or higher strata) cannot somehow somatize. In other words, whether or not what is originally inherited culturally (only through social learning) cannot pass into the plane of epigenetic, physical inheritance (e.g. brain settings). This is a wonderful speculation, and some indications might suggest it, but we will refrain from taking this step—although we did in fact move in this direction in Sect. 6.6.
2. Let us try to define culture as a *sign system of relations* that accompanies life *from the very beginning*. Everything that can be called culture fits the definition of "life" as a bodily, born semiotic system with history. A different approach says that not every semiotic system is culture, and at some point, there was a *separation of higher, more autonomous layers of culture* from bodily inheritance, and they began to spread in ways other than "life itself". These later layers of culture then represent a certain *system of meaning* that is *(semi)autonomous* with respect to life. The reader already knows that we are inclined to the first alternative. It can perhaps be stated that biosemiotics returns *all critters* with their innerness (in which processes of semiosis can take place) to the mainstream of life evolution.
3. There is an extremely interesting question of the "chance" that a certain interpretation—whether it takes place "ontically", in all living beings, or is made by humans, i.e., biologists or historians—will reach its intelligibility, acceptability, or just tolerability within a certain "field" (ecosystem or socio–cultural field, or culture–space; this does not mean that it should reach general sharing). We humans can perhaps to some extent retrospectively distinguish those (theoretical, narrative) interpretations that "succeed" from those that failed. In terms of scholarly interpretations, it is easier, it's simply a matter of the evolution of the scientific debate. But what about interpretations made by living beings? We leave the question hanging in the air; further research will surely yield something along these lines.
4. Are the factors of the development of *Umwelt* or "the world" immanent to life, or is it an adaptive response to external pressures only?
5. Can "evolution" and "history" be distinguished from each other in an *absolute sense*, or will the answer always depend on our starting point chosen? Again, there is an extensive literature on this point and there does not seem to be a clear–cut answer. For explanatory reasons, we have in this book stuck with the axiom that history and evolutionary biology are analogous approaches to the past, which, in fact "no longer exists".

Alongside the innerness and semiotic endowment, the second neuralgic point towards which discussions about the history of the living often turn is the *driving force* of the process itself. It is argued that while the evolution of non-human beings is driven by an external *natural* force—natural selection, the driving force of history is only humanity. (We will not discuss the third possibility, creative intervention from the transcendent realm, as this would require evidentiary support beyond our current capacity to ascertain.) This split perhaps came about during the second half of the nineteenth century, when historians largely rejected historicism, and biology became even more "mechanistic" and determined to derive life processes from physics alone. Such a schedule, of course, again precludes biosemiotic considerations—but that is not of interest at this point.

Let us stop at the assumption of the physical nature of natural selection. For it is itself—as we have shown above—based on an analogy with a Newtonian force, namely gravity, the invisible hand directing the swarming of a large numbers of particles within its reach. Darwin himself—as far as we know—nowhere says this explicitly, but he *was* an adherent of Newtonian physics, and his doctrine originated as an attempt to reconcile timeless physical laws with the fact of evolution. His followers then elaborated the approach, and for many, the dominant role of the evolution of life (as opposed to the history of mankind) is still attributed to physical factors: on the one hand, natural selection, on the other hand, chance, which doesn't even have to be a chance, as it is the result of physical forces summarized under the name of quantum mechanics. This assumption seems to prevent both camps—historians and evolutionists—from seeing that both historical processes are analogous, indeed dare we say homologous. Both are based on cells that, from our perspective, automatically presupposes two kinds of Closure, and, above all, on the innerness and the resulting intentionality and capacity of the living to manage its affairs, be it the affairs of cells, species, or large heterogeneous groups of organisms, *including* humans. In this book, we attempt to unify both processes, let us call it a *General Model of Historiography*, though the determination of its contours will require further work.

Closure 1 concerns the very basis of life—the cell and its functioning, reproduction and perception of the world around it (which is of course mainly represented by other life forms). We have shown (Markoš & Švorcová, 2019) that there are certain norms remaining unchanged (or changing only slightly) in all lineages of life, hence they can be assumed to have been established at the very beginning, i.e. in the LUCA biosphere. Consider a biospheric "global internet", for example in the form of horizontal gene transfer, and probably some generally established signals and rules. It is true that they could have emerged later by the joint efforts of the inhabitants of the biosphere, and then continuously maintained; as an analogy, we can then consider the global development of communication techniques such as the http protocol, which had no "primitive" form 100 years ago, and yet are now shared by the whole of humanity. However, we have good reason to believe that such global networks have existed in the biosphere since the beginning. But alas, the sources are rare, and their interpretation difficult!

As the different lineages of life separated over the ages, various modifications and extensions to the original norms became established, and of course new norms were introduced, now maintained only as part of the evolution of one of the lineages. Care also had to be taken to limit the "informational promiscuity" between them. What is important, however, is that between the lineages there is nevertheless at least minimal overlap in the understanding of the modified norms, and over time—and with longer cohabitation—some understanding of the new norms emerges. Simply one must *understand* the changes and novelties in the environment, or one will likely perish!

While Closure 1 can be studied at the cellular level, Closure 2—i.e. the behavior of cellular communities—is approached on the one hand by the multi-layered nature of such communities (Fig. 11.1), and on the other hand by the fact that a single cell or community can simultaneously belong to different communities of a Closure 2 community, and can change its affinity for different groupings according to external events. But Closures 2 have also been with us from the earliest days of life and have contributed in no small measure to evolution. We emphasize again and again that evolution is a property of the mutual elicitation of Appearances that are *relevant* to the moment. Even more precisely: Appearances that are *interpreted* as relevant at a given moment.

Let us ask once again: how can we grasp the common characteristics of such diverse and multistage "associations" as, for example, cooperation between two types of bacterial cells, a multicellular organism, a football fan club, or the citizens of Switzerland? Undoubtedly, some form of reduction must be applied: in biology, the more common and methodologically graspable approach is "bottom-up", starting from the properties of cells and their genes. It is a legitimate approach, and many properties of even complex assemblages can be explained, for example, genetically (i.e., from cell type differentiation to ethology). At the genetic level, heritable and congenital disorders affect even complex cellular assemblies such as human bodies and their psyche. Biologists, on the other hand, look with suspicion at "anthropocentric" approaches to the living, as if the human approach to the phenomena of the world were completely outside the knowledge of how things "really" are. Yet the approach from above (from the human being) could also be used as a (more or less) unified explanatory scheme for all types of groupings and their histories. This is precisely the possibility we are looking for in our text, which is why we have chosen analogies to analyze what is closest to us: the history of human communities, the affinity of individuals and groups and of individuals to certain groupings, the affinity to common memory, experience and interpretation of the past, and its changes over time, including estimations of the course of things to come. In addition, humans have at their disposal language and thought, ritual and tradition, externalized memory in the form of writing, etc.; and yet, despite all this, one could perhaps detect characteristics and practices independent of such human apomorphies. What is common can be found at the level of *umwelt*, admittedly in our own interpretation of the term. In fact, this is why we, the authors, joined forces to write this book—we wanted to illuminate both extremes, hence for Closure 2 we started from our human experience and tried to work our way back (top-down) to the primordial ancestor of both Closures.

If we accept the premise that all types of Closure 2 in human communities are also reflections of biological evolution, we can look for such Closures 2 in humans that *do not* necessarily depend on the existence of apomorphies such as mind and speech. Examples might be found in the taboo of incest, in the competition for sexual companionship, or in sexual selection and its associated fashions. Even more biologizing are the "artificial ecological relations" in the form of domestication of plants, animals and microorganisms, which is ecological because domestication is a *mutual* process, it is impossible to point out who domesticated whom—the process involves humans as well. If we accept this, we are shown biological Closures 2 of shorter duration and reach, reflecting human situations "here and now", as well as their dynamics. By this we mean, for example, the loosening or changes of emphasis within the cellular communities that form the individual; or changes in ontogeny leading to increased occurrence (or new establishment) of manifestations such as tumors, allergies, changes in psyche, and new ways in the setting of holobiotic relationships; or last but not least, raising children or ignoring or exploiting key developmental time windows in established ontogenetic processes.

We can then try to extrapolate from this to non-human life forms and look through this lens, for example, at the self-expression of organisms within a species; and again, at the *umwelt*, speciation, and ecological relationships of different species, epidemics or invasions, etc. In all such examples, it is about how a community (whether of cells or organisms) perceives certain kinds of Closures—based on a "perception of the self" confronted with an interpretation of signs coming from outside, all filtered through perception of the past of the individual and its community. Perhaps this can be summed up by the slogan "that's how we've always done it here", or, in contrast, by discarding such a slogan and attempting abduction in newly emerging historical moments.

We repeat—we try to extrapolate from human experience to all living things. Through memory and experience we can awaken ancient patterns of Appearances and place them in new contexts. Different atavisms persist in virtual form in both biological and cultural evolution: sometimes they appear by chance and then disappear because they represent anachronisms; at other times, it is precisely by reviving ancient (more primitive?) ways of solving problems that one can "start over", or try doing things differently, and restore plasticity as an antidote to the above-mentioned hardened, frozen slogans. Doing things differently—even with all the risks that may entail.

At the end, some outcries similar to those formulated in Markoš (2026) as a chain of Yes and No statements (paraphrased):

No, to physicalization of life. Hylozoic, panpsychic, physiosemiotic and similar activies may provide insights to the dynamics of our universe, but cannot be extended to living beings.

Yes, to a belief that life is a special quality in the universe.

No, for external cues operating on life as if "clicking" on a computer icon. Hence:

Yes, for the view of codes (norms, habits) negotiated in evolution of life, and life only, but.

No to their rigidity set in advance, reducibility to a kind of entries in Excel tables.

Yes, knowledge developed by humans as living beings, should be taken as an analogy for extending it to many of life's doings. Therefore:

Yes to the biosemiotic endeavor.

References

Danchin, É. (2013). Avatars of information: towards an inclusive evolutionary synthesis. *Trends in Ecology and Evolution, 28*, 351–358.

Deely, J. (2010). *Medieval philosophy redefined Scranton*. University of Scranton Press.

Depew, D. J., & Weber, B. H. (1996). *Darwinism evolving. Systems dynamics and the genealogy of natural selection*. MIT Press.

Flegr, J. (2010). Elastic, not plastic species: Frozen plasticity theory and the origin of adaptive evolution in sexually reproducing organisms. *Biology Direct, 5*, 2.

Flegr, J. (2021). *Frozen evolution. A farewell to selfish gene*. Andrias Press.

Futuyma, D. J. (2021). How does phenotypic plasticity fit into evolutionary theory? In D. W. Pfennig (Ed.), *Phenotypic plasticity and evolution. Causes, consequences, controversies* (pp. 349–366). CRC Press.

Gurevich, A. (1990). *Medieval popular culture*. Cambridge University Press.

Lotman, Y. M. (2009). *[1992] Culture and explosion*. Mouton de Gruyter.

Lotman, J. M. (2010). *Semiosfera*. Iskusstvo – SPB.

Markoš, A., (2026). Taking life stance. In K. Kull & D. Favareau (Eds.), *Toward a Biosemiotic Theoretical Biology. Sign Processes and Meaning-Making in Living Systems*. MIT Press, 253–266.

Markoš, A. & Švorcová, J. (2019). *Epigenetic processes and the evolution of life*. CRC Press.

Pfennig, D. W. (Ed.). (2021). *Phenotypic plasticity and evolution. Causes, consequences, controversies*. CRC Press.

Rappaport, R. A. (2010). *[1999]. Ritual and religion in the making of humanity*. Cambridge University Press.

Švorcová, J., & Markoš, A. (2023). Closures as a Precondition of Life, Agency, and Semiosis. *Biosemiotics, 16*, 45–59.

Toman, J. & Flegr, J. (2018). Macroevolutionary freezing and the Janussian nature of evolvability. Is the evolution (of profound biological novelty) going to end? *Biosemiotics, 11*, 1–23.

Epilogue

'After all, is it any business of mine how things are in reality? My task is to invent, isn't it, to play, pretend– 'The poet leaned forward with his heavy shoulders. 'That's just the trouble, sir: reality means such a frightful lot to me. That's why I invent it, that's why I always have to invent something to catch hold of it. What I see with my eyes is not enough, I want to see more—and so I invent stories. Please tell me, is there any sense in it? Has it anything to do with life at all? [...] I know that it's ... only a fiction. And I know, my friend, what fiction is, I know how it's done: one part experience, three parts phantasy, two parts logical construction, and the rest artful guile: that it is fresh, that it is topical, that something is being solved or proved in it, and chiefly that it's effective. But one thing is peculiar.' the poet burst out, 'that all these tricks, all that miserable literary hackwork leaves the man who performs it with an accused and passionate illusion that it has something to do with reality.

Karel Čapek, 1990, p. 186

A. Markoš, J. Horský, *The Seeds of Forms and the Forms of Seeds*, Biosemiotics 12, https://doi.org/10.1007/978-3-032-16131-4

Summary

1. Two themes are intertwined in this text: historiography and evolutionary biology. The aim is to show that both activities are analogous in their practices—they are spinning stories about the past and thus constructing the world to suit the present of a given community as well as, if necessary, the construction of its tilting into the future. This text presents the history and methodology of both branches of historical narrative and demonstrates them through case studies.
2. The theme of both branches of history is the interpretation of the history of life, its evolution on the planet. For methodological reasons, life activities are presented here as manifestations of two Closures: the cellular "Closure 1" and the supracellular "Closure 2". The dynamics and interdependences of the two Closures is the focus of the authors' attention.
3. All life is cellular, and today's cells—representing Closure 1—are direct descendants of the primordial cells as they were constituted in the singularity of life's origin. The individual lineages of life preserve the norms established in the beginning, and—in course of evolution—accumulate new variants and novelties contributing to memory and experience traces that help them interpret the world.
4. Supracellular forms of Closure 2 represent organized communities of cells as well as, recurrently, a network of manifold of such Closures 2. They arise and may cease to exist after some time, and necessarily require—in the background, the activities of Closure 1. This ability to constitute a variety of ecological relationships (cultures)—from simple cellular consortia to communities of organisms (including human civilizations) to the biosphere—also has its basis in the primordial singularity, but given the emergence and disappearance of its various forms, its manifestations are historically contingent.
5. For exploring both branches of historiography, the tool is the discovery and interpretation of historical sources (traces) and their integration into a coherent narrative. These are thus characterized by the interplay of scientific and (bio) semiotic approaches built on the infrastructure of a given culture (scientific instruments, archives and political milieu, but also climate, biome composition

A. Markoš, J. Horský, *The Seeds of Forms and the Forms of Seeds*, Biosemiotics 12, https://doi.org/10.1007/978-3-032-16131-4

etc.). The narrative—and thus the image of the world that a community constructs—is therefore fluid, dependant on the memory and experience of the community but also on the community infrastructure.

6. Alongside the activity of historiographers and evolutionary biologists, the question arises whether individual lineages also constitute their world in analogous ways. This book provides a brief characterization of the biosemiotics that seeks to answer the question.

Glossary

Abduction A hypothesis (qualified guess or opinion) about the nature (origin) of an observed phenomenon or event that cannot be derived by logical operations such as deduction or induction.

Adaptive landscape In genetically heterogeneous populations, different genotypes may provide a comparable phenotype. It is therefore customary to compare fitness rates of different genotypes on a graph with adaptive peaks (highest fitness) and valleys in between (maladaptation). Natural selection then determines the position of a given genotype on the slopes, peaks and valleys of the landscape in any given environment; the position relative to the peak represents the fitness level of the phenotype. (The landscape is multidimensional; the one-dimensional expression in the graph below evaluates only a certain genotype-fitness or allele-fitness combination.) A population on one of the "hills" either climbs to the top over generations, or falls back a bit, and only rarely may jump across the "valley of maladaptation" to another of the neighboring hills. As Kauffman (1993) points out, the landscape itself is a product of evolution: a landscape of the shape of a gradual hill suits evolution, whereas a the "column in the plain" type is rather unthinkable. It must also be assumed that the relief will not change faster than the distribution of alleles in the population.

Allele In this text, a variant gene in a diploid or polyploid genome (see *ploidy*). In a diploid genome, the identity of the two alleles defines a homozygote; two different alleles define a heterozygote. The alleles may differ in their expression, i.e., the way in which they affect the function or form of a cell or multicellular body. Their manifestations may vary in a non-trivial way (e.g., a combination of alleles coding for white and red genes does not always result in a pink flower), and of course also affect the expression of other genes located in other places in the genome (see epistasis). In a population there may exist a larger number of alleles, but a diploid individual may acquire only one or two from them.

Analogy Similarity recognized from a certain viewpoint; one of the key concepts in our reasoning, interconnected with the entries like *Appearance, phenotype*, or

A. Markoš, J. Horský, *The Seeds of Forms and the Forms of Seeds*, Biosemiotics 12, https://doi.org/10.1007/978-3-032-16131-4

homology. Caveat: in some way or other, everything shares some similarity with something else (trivial examples: the mass of bodies or metaphors like "slave labor") - so the viewpoint of recognizing analogies is important. If a creature is characterized by its form (Appearance in our text), its individual characteristics can be expressed as a phenotype; the phenotypes can then be compared and declared analogous in totally unrelated bearers. Examples: the wings of a beetle and those of a bird, the achenes of a maple tree, and even the wings of an airplane can be said to be analogous structures designed to move their bearer through the air. But in no case is it an identity. But the matter is more complicated; let us first try with mathematical examples. (1) The trinomial is certainly a good tool: if we know the relation of two quantities (a, b) and the value of the quantity in the same relation (c), we can easily calculate the value of the fourth (x; $a/b = c/x$). We can calculate, for example, the amount of sand needed for plaster or the area of carpet in a hotel corridor. Of course, the rule is the same, but what is the analogy between these quantities? Probably we can agree that there is none. (2) Similarly, the formula for calculating statistical entropy (Boltzmann) is identical to the calculation of statistical information (Shannon), only with the opposite sign. (3) Let us now turn to life with its history. The wings mentioned above are certainly analogous, but if we exclude the airplane, a problem arises. The wings of the bat and the bird are certainly analogous, but from the point of view of vertebrate evolution, they represent the forelimb of quadrupeds, i.e., they are homologs. Let us now turn to human history and ask whether also manifestations of life such as various forms of "oriental despotism" or "slavery" are analogous, i.e., did they develope independently or were established ad hoc (then they are analogies), or are they a product of evolution from some ancestral form of society (then homologies), or even are they are determined by "natural laws" in which case they represent an unchangeable, teleological, scientifically describable process. We call upon the reader to stick to the idea of similarity in our text.

Apomixis The sexual process need not represent only the classical fusion of two sex cells. In apomixis, the egg is produced from a somatic cell without meiosis, i.e., the new individuals are a clone of the parent plant; an example is dandelions.

Apomorphy A new characteristic typical only of a given lineage or life – i.e., not present in closely related lineages. Contrast *plesiomorphy*: a trait shared by several distinct but related lineages.

Appearance (with a capital *A* here: also habitus, eidos, form, Gestalt). The overall phenomenon of a creature, as co-shaped by genetic influences and environment, moreover, differently in different stages of ontogeny. Unlike the *genotype*, which can be "read" objectively, as a sequence of discrete units, the Appearance cannot be defined in this way, hence its frequent reduction to *phenotype*.

Bohemia syn. Czech lands. Today the western part of the country called Czechia or Czech Republic.

Cell The basic and inherent structure of all life, characterized by continuous existence into the past; in this text, *Closure 1*. In today's biosphere we find three types of cells: bacterial, archeal and eukaryotic. The first two types are represented by the smaller size (1–2 μm in diameter) and have historically been referred to as prokaryote; eukaryotic cells are 10 μm or more in diameter and are characterized by the presence of a cell nucleus. It is thought that eukaryotic cells originated symbiogenetically, by fusion of cells from the first two groups. The biosphere is composed of a variety of formations built by these three types of cells, referred to in this text as a *Closure 2*. Unlike cells, such formations can arise and disappear again.

Central dogma of molecular biology Genetic information is transmitted from nucleic acids to proteins and never the other way around. Proteins then take care of everything else in the "construction" of the body. Thus, from this, all life processes and structures are believed to be encoded in the genetic transcript. The *fluctuation test* is considered a strong presumption of confirmation of the dogma, *epigenetics*, on the other hand, significantly weakens its validity.

Closure For an explanation, see Introduction; see also *innerness*.

Code Very briefly, a guide for transcribing information from one medium into another, e.g. from the spoken word into writing or vice versa. See also *genetic code*.

Coinoscopy A doctrine of knowledge based on speculative, interpretative knowledge.

Contingency Unlike mere chance, contingency allows us to grasp a random event, it is "chance grasped creatively". For example, if during an earthquake drops a ceiling on your head and kills you, it's a coincidence and you can hardly do anything about it. But if you manage to run out into the street, the event may be taken as a challenge to thriving – and building a better and more beautiful house.

Continuity of evolutionary (developmental) processes According to most researchers in the humanities, historical process may be related to the fact that the development of culture, unlike the development of life, does not exhibit a physical continuity, or at least not as strongly as the evolution of life. Here, we define life as a "bodily, born, semiotic system with a history" (Markoš, 2016), and so we include "life" in the general sense of both biological species and (human) cultures, as subdivided into two types of *Closures*.

Culture Here, we work with the concept of culture primarily as a semiotic system (i.e., a system of meanings) that is disseminated through social learning. In a broader sense, we can also say that culture is a system of meanings, values and patterns of behavior or action (semiotic, axiological and praxeological) and that we understand social learning very broadly to include the dissemination and sharing of meanings within individual Closures 2. From a different perspective, we are also working with culture as a system of cultural complexes that consist of cultural variants.

Czech The language and an ethnonym of a Slavic ethnic group inhabiting *Bohemia*. Until the middle of 1940s, a second ethnic group was German.

DNA A linear molecule (a non-random polymer of four so-called bases) storing genetic information through a sequence of bases (i.e., the "script").

Elasticity The idea that established lifelines resist the action of external environment. After such a 'selective' action ceases, the Appearances (overall phenomena of creatures) return to the previous way of life.

Enhancer A stretch of DNA that can bind regulatory proteins; upon binding, genes can be turned on, off, or be activity modified

Entelechy In vitalist doctrine, the hypothetical force required to induce and drive life processes.

Entropy In physics, a statistical quantity expressing the degree of spontaneous decay of the orderliness of the world.

Epigenesis Development of the body (or better, *Closure 2)* from simple beginnings (e.g., embryo from egg, adult from larva, or a holobiont from newborn). Conditioned by the interplay of genetic information, the previous steps of epigenesis and environmental influence.

Epigenetics The science dealing with various aspects of *epigenesis*, from molecular processes to the formation of the Appearance. In this text, esp. the controlled chemical modification of biological macromolecules (*proteins, DNA, RNA* and *polysaccharides*).

Epistasis The interaction of the products of many genes (i.e., RNAs and proteins) to produce a phenotypic trait.

Eukaryote see *cell.*

Family estate This refers to the law of property and inheritance in which an estate is owned and inherited by the whole family. If one of the men wants to leave, he leaves with the corresponding share of land. In contrast to the so-called single heir system, this system did not permanently separate part of the rural population from the land (and thus did not generate landlessness).

Fitness The statistical rate at which the bearer of a phenotype will reproduce that phenotype into subsequent generations.

Fluctuation test Testing the hypothesis that mutations in genetic material are random, not environmentally induced. The result led to the development of the mutation-selection model of evolution, still widely accepted today.

Gene 1. An endowment determining a *phenotypic* trait. 2. In a narrower sense, a section DNA that is transcribed (*transcription*) into RNA. 3. In an even narrower sense, a stretch of DNA that is transcribed into RNA, which is then edited into mRNA and encodes a protein (*translation*). Here we usually use the term in the meaning 3.

General history In the epistemological sense, the discipline of historical sciences that seeks (usually by means of comparative methods) to uncover certain invariants that occur along multiple lines of historical development, and at the same time describe the specifics of each variant in relation to these invariant patterns. The legitimacy of this discipline is debated today. Where it relies on the possibility of revealing the content of general history in an ontological sense, the practices of scholars approach an attempt to uncover the historical laws in the spirit of *historicism*.

Genetic assimilation The path from phenotype to the altered genetic information. An opposite, often only apparent, to the central dogma.

Genetic code A prescription for the synthesis of *proteins* as polymers of 20 amino acids according to a chain of mRNA. Each amino acid is encoded by one or more (up to six) consecutive triplet bases in the mRNA. The reading frame (there are three possible) is set by the translational structures of the ribosome.

Genetic notation A non-random sequence of elementary components (bases) in a linear polymer of nucleic acids (DNA or RNA).

Genetics The science of the inheritance of phenotype; nowadays mostly with the addition of "by genetic transcription in DNA".

Genome is the way all the DNA in a cell is organized. For example, a human has about three billion bases (about one meter of strand) in one genome set; these are broken down and stored in 23 discrete structures ("chapters") - chromosomes; we carry two unique sets, one from each parent. Different versions of the "chapters" of the genome thus make it into each generation variable contexts with other chapters. In the DNA chain, it is usually possible to distinguish well-separated sections - let's call them *genes*, and these are characterized by their precise coordinates - a map or address (e.g. the X gene is located on chromosome 11, its q arm, at 34567-40100 bases from the beginning of the chain).

Genotype A version of the genome.

Historicism, historism Historicism refers to the belief of the possibility of finding a basic pattern in the general course of human of history. Such a pattern can be used to make causal interpretations of changes of individual historical formations and, where appropriate (according to some authors) to predict the next development. By historism, then, we mean the belief, that historical development is a constant teleological diversification, culminating into the peculiarity of individual historical formations.

Historicist nominalism presupposes that theoretical constructions of historical processes or their historical narrative interpretations do not, or may not have an (exact) ontic counterpart.

Homology The relationship of genealogical relatedness of certain entities. In biology, the refers to structures (e.g., the forelimb of vertebrates) or genes (orthology, see nalogy.

Hylozoism (also Hylonomism). The idea that the world and *all* its parts (i.e.. not only its parts described here as *Closures*) are endowed with life.

Character 1. qualitative or quantitative criterion for evaluating a phenotype. 2. In semiotics, a suprasubjective reference to meaning.

Chromatin A dynamic structure containing most of the genetic thesaurus of the cell, organized and controlled by thousands of proteins and RNA. In eukaryotic cells, chromatin is more or less synonymous with of the cell nucleus.

Chromosome A section of chromatin. Human chromatin is divided into 46 chromosomes.

Ideal type One of the central concepts of Weber's (1904) cultural noetics, of the noetics of the historical and social sciences. The ideal type is a conceptual tool knowledge, precisely defined *a priori* with respect to the topic under investigation (i.e., it is not defined *a posteriori* by inductive generalization of the observed phenomena). The researcher determines how the studied reality is close or distant to the respective ideal type.

Idioscopy The science of knowledge based on investigations of modern (experimental, natural) sciences.

Imbred population Individuals in a population are inbred with each other (for example, father-daughter) so that after about 50 generations they exhibit the same genotype (except for sex differences).

Information Coincidentally, the formula for calculating physical entropy (S) and Shannon's information (-S) agree, except for the sign. Many conclude that the two quantities are on a single scale representing the degree of complexity, the complexity of the thing, the world, anything. Whether rightly or wrongly, we leave it to the experts or the opinion of the reader. In any case, it is a reduction of the world to an extreme degree, where meanings and their interpretations go by the wayside. Information so conceived can only get lost in the system (dissipate into noise, entropy), it can never emerge.

Initial and boundary conditions Important prerequisites for any scientific experiment. In historical investigations, the initial conditions can never be accurately determined because, in principle, it is impossible to accurately map the history, memory and experience of individuals, populations, societies. In biology, therefore, the tendency is to work over the long term with homogeneous clones or populations of organisms isolated from the outside world and also kept in stable conditions for long periods of time. In describing (or calculating) the rate of movement of a body from the gallery of a tower to its base, it is necessary to know in advance, for example, the height of the tower, the mass and volume of the body to be examined, or the atmospheric pressure and the elevation of the gallery; boundary conditions are, for example the method of movement of the body (free fall, descent down the slope of certain parameters, rolling down stairs), the shape of the body (sphere, box, irregular stone, feather). For historical examination, the initial conditions can never be precisely determined because, in principle, it is impossible to map the history, memory and experience of individuals or populations. If we look for analogies between the evolution of life and the evolution of culture, we can probably establish such conditions only in a particular section of the period. The question is whether the distinction between biology and history is principal at this point, or whether it is merely a difference on an empirical level, or interpretative certainty. If the latter is true, then the question is, to what extent the possibility of distinguishing between "evolution" and "history" is absolute, or only by the nature of the research perspectives chosen.

Innerness The ability of living creatures to represent (to themselves) the external world on the basis of their own memory and experience, and of course also according to the momentary context, to act. Both forms of *closure*, and the historical continuity of life, are a prerequisite. Memory and experience then originate in the bodily structures (including discrete storage in DNA) and the evaluation of inputs from the world based on their own experience, the experience of a given lineage or group (community).

Meaning The degree of importance that a living entity attributes to a certain thing or process, according to the current situation (context), according to one's own memory and experience.

Metaphor A way how we conceptualize one mental domain in terms of another. Basis for all forms of *analogy*.

Mimesis Imagery of unrelated life forms unrelated to environmental features, e.g., cover, warning, or threatening coloration.

Mutation A change in information, process, or structure. In the context of this of this book, an irreversible random change in the sequence of bases in the nucleic acids (substitution, deletion, or insertion) that may result in a change in the sequence of amino acids in a protein, and thus may cause change the phenotype of the cell or of the supracellular communities.

Mutation-selection model of evolution. Random mutations of genetic material (e.g. the emergence of a new allele) will cause a change in phenotype. Selection (natural selection) then - in the context of internal and external contexts - determines the fitness of the mutant.

Natural selection In Darwinian models, the principal factor "driving" evolution. *Phenotypes* are selected according to their *fitness,*

"Noetic" and "ontological" perspectives in the historical sciences. Where we speak of the history of human culture and of historical science and its approaches, we often distinguish both approaches. From the point of view of biology, this distinction may appear as redundant, even confusing. Yet we witness disputes as to whether, e.g. biological species exist in reality - not to mention all levels of taxa higher ones.

Norm The establishment of codes, arrangements, and modes of behavior that is (with variations) observed in most lines of life (culture).

Object (semiotics). A certain universally accepted and teachable aspect of knowledge of a thing; also, in biology, a phenotype. The "unknown object" in this sense of oxymoron.

Organoid An in vitro, cell-culture-grown formation resembling the structure of organs (e.g., stomach, forebrain, or early embryo).

Ortholog A gene that also has a counterpart in the genome of other creatures; see also homology.

Palimpsest Ancient text, structure, or behavior reconstructed from *characters* that were superimposed by more recent variants.

Paralog A variant gene that has related form(s) in the same genome. Example: variants of the FOXP gene.

Phenotype Body trait—a structure or behavior; one of many aspects of *Appearance*.

Plasticity *1.* The ability of an individual or population to respond to change conditions by changing its phenotype (e.g., hairy plants in cold climates; winter forms of some species); 2. Flegr (2010): The first, short phase of a species' evolution characterized by the ability to change form in different directions due to conditions, and then to establish a subsequent (elastic) trajectory of evolution.

Plesiomorphy see *apomorphy*.

Ploidy A multiple of set of in the genome of a cell; a single set of a genome defines a haploid cell (in humans, only sex cells and their immediate precursors); similarly, two sets define a diploid cell (e.g., most human body cells); in plants, polyploidy is common.

Polysaccharide A linear or branched polymer of a small set of simple sugars; the order of sugars in the polymer is a notation, a carrier of information to the receptive cell. There are also storage polysaccharides - as starch, inulin or glycogen -, or building - like cellulose; these are not dealt with in this text.

Prokaryote see *cell.*

Protein A linear polymer of a small number (usually 20) of amino acids. The chain folds into a three-dimensional dynamic structure performing a specific function.

Protista Single–celled eukaryotes.

Quasi–character Any Closure 2 that exhibits the characteristics of an individual, i.e., "wants", "demands", "controls", etc. E.g., a nation, a human body, a state.

Representation (see also *innerness*, see also *umwelt*). The ability of a living thing to make itself to make sense of the world by evaluating memory, experience and contexts internal and external (e.g. ontogenetic stage, season, the presence of other life forms, etc.). Often the term is linked to linguistic abilities, but the examples in the previous sentence also draw attention to its non-linguistic forms. In the (narrativist) theory of history metaphorical representation is sometimes contrasted with to object-relative factual reference. At the same time, it is also debated whether (theoretical and narrative) historiographical interpretations can mirror past reality or not.

Ribosome The structure of the cell carrying out *translation* of genetic information into a *protein* according to the prescription of the *genetic code.*

RNA A linear molecule (a non-random polymer of four so-called bases) that is created by copying (transcription) sections of DNA (*gene* in the meaning 2), thus representing a copy of a part of the genetic transcript. Of the many functional of functional types, our text lists three types of RNA that are involved in protein synthesis in the process of *translation*: mRNA is a unique prescription for translation itself, rRNA and tRNA are the molecules that enable such a translation. In addition, there are several types of regulatory RNAs, see especially Chap. 10.

Sacraments Sixteenth-century theology (let us leave aside here the contemporary theological conceptions), defines sacraments as visible signs of the invisible grace of God. From the point of view of anthropology or sociology, we can say that the sacraments are ceremonial ritual acts with specific value-attachments and beliefs, or theological concepts. In the individual denominations of Christianity, different numbers of sacraments are celebrated, and the doctrine of their functionality may also differ

Semiosis The interpretative method of *semiotics.*

Semiotics A science describing ways of interpreting phenomena of the world. In this book the focus is on semiotics as elaborated by Charles S. Peirce and his followers from the turn of the 20th-21st century - Umberto Eco and John Deely.

Sources Geological, paleontological, archaeological, written, witnesses of the past. Also, entities that can be inferred from the current state of the community (e.g. cultural behavioral patterns or comparative analysis of different life lines or culture), such as those that allow the reconstruction of historical events. Whether something is a source depends on the agreement of the community of researchers, so the status as a source may not be permanent.

Teleology The doctrine that the course of the world is directed towards a goal—the revealed or knowable by one's own efforts.

Thing According to J. Deely (1) An entity in the world existing independently of our consciousness. To know it, we construct (by semiotic methods) *objects*. (2) An entity that arose in the mind and whose properties we recognize by the semiotic construction of *objects*.

Transcendental philosophy of history One of the two modes of the philosophy of history (along with the content-based, addressing questions such as "the meaning of history", etc.).

Transcription The transcription of a stretch of DNA into a strand of RNA. Translation. Synthesis of a protein molecule according to the information carried by mRNA, according to the genetic code set by the ribosome.

Author Index

A. Markoš, J. Horský, *The Seeds of Forms and the Forms of Seeds*, Biosemiotics 12, https://doi.org/10.1007/978-3-032-16131-4

Subject Index

A. Markoš, J. Horský, *The Seeds of Forms and the Forms of Seeds*, Biosemiotics 12, https://doi.org/10.1007/978-3-032-16131-4

MIX
Papier aus verantwortungsvollen Quellen
Paper from responsible sources
FSC® C105338

If you have any concerns about our products,
you can contact us on
ProductSafety@springernature.com

In case Publisher is established outside the EU,
the EU authorized representative is:
Springer Nature Customer Service Center GmbH
Europaplatz 3, 69115 Heidelberg, Germany

Printed by Libri Plureos GmbH
in Hamburg, Germany